SpringerBriefs in Mathematics

SpringerBriefs in Mathematics showcases expositions in all areas of mathematics and applied mathematics. Manuscripts presenting new results or a single new result in a classical field, new field, or an emerging topic, applications, or bridges between new results and already published works, are encouraged. The series is intended for mathematicians and applied mathematicians.

For further volumes:
http://www.springer.com/series/10030

W.T. Ingram

An Introduction to Inverse Limits with Set-valued Functions

 Springer

W.T. Ingram
Professor Emeritus
Department of Mathematics and Statistics
Missouri University of Science
    and Technology
1870 Miner Circle
Missouri, USA

ISSN 2191-8198   ISSN 2191-8201 (electronic)
ISBN 978-1-4614-4486-2   ISBN 978-1-4614-4487-9 (eBook)
DOI 10.1007/978-1-4614-4487-9
Springer New York Heidelberg Dordrecht London

Library of Congress Control Number: 2012940256

Mathematics Subject Classification (2010): 54F15, 37B45, 54H20, 37E05

Printed on acid-free paper

Springer is part of Springer Science+Business Media (www.springer.com)

To my wife, Barbara, our sons, Bill and Mark, and our daughter, Kathie

This work is also dedicated to the memory of Bill Mahavier who got us started thinking about inverse limits with set-valued functions and to the memory of my advisor, Howard Cook, whose influence on my career is immeasurable.

Preface

Inverse limits with set-valued functions were introduced in 2004 by Bill Mahavier as inverse limits with closed subsets of the unit square. In the short time since then, the subject has rapidly developed into a rich topic of research, particularly among continuum theorists. This new form of inverse limit can also show up in applications to economics and in dynamical systems. For instance, certain models in economics, notably in backward economics, can involve two mappings, and the flexibility to study the effects of using either function at each stage in the model is a valuable feature of inverse limits with set-valued functions. Recent work of Benjamin Marlin on the Lorenz attractor has provided evidence of the value of using set-valued functions in studying attractors in dynamical systems. This short book is not specifically concerned with these applications of set-valued functions. Instead, it is intended to provide a quick introduction to the subject of inverse limits with set-valued functions. The recently published Springer book *Inverse Limits: From Continua to Chaos* contains most of the background a researcher might need to make use of these inverse limits in his or her work. However, a shorter volume such as this one dedicated to inverse limits with set-valued functions could be helpful to someone wanting a quick introduction to this specific subject. This short book is intended to provide just such an introduction. The emphasis here is on inverse limits on the interval [0, 1] much as the first chapter of *Inverse Limits: From Continua to Chaos* serves as an introduction to inverse limits on [0, 1] with mappings. We believe that a deeper understanding of inverse limits can be obtained by studying examples. A major feature of this book is the inclusion of numerous examples and, in many instances, models of the inverse limits. Several of the examples in this volume have not appeared elsewhere in print.

Major differences between the theory of inverse limits with mappings and the theory with set-valued functions arise early in this study. These differences are featured prominently in this book. However, instead of viewing these differences as a negative development, we consider them as an opportunity for further study and research. Indeed, much of the research in the subject is devoted to resolving questions arising for these very reasons. We have included an extensive, but by no means exhaustive, list of currently unsolved problems in the final chapter of the book.

A senior-level course in analysis and, perhaps, an additional one in topology should provide a sufficient introduction to the topology of metric spaces and the topology of product spaces to make the material in this book accessible to advanced undergraduates and, certainly, to graduate students in mathematics. An alternate source of background material is the appendix in the Springer book *Inverse Limits: From Continua to Chaos*. In some of the examples, we assume some familiarity with inverse limits with mappings, but no deep understanding of ordinary inverse limits is necessary to read this book.

This book is based on a series of lectures given by the author at a workshop in the summer of 2011 at the Instituto de Matemáticas, Universidad Nacional Autónoma de México, in Mexico City. Many thanks go to all who participated in that short course. We are particularly indebted to Verónica Martínez de la Vega and Alejandro Illanes who organized the workshop and helped make our stay in Mexico City such a pleasant experience.

Spring Branch, Texas   Tom Ingram

Acknowledgments

The author appreciates the assistance of Van Nall, Michel Smith, and Scott Varagona who read significant portions of the manuscript. Without their proofreading skills, there would have been many additional typographical errors in this book. The author remains solely responsible for the errors that remain.

The author appreciates the assistance and support of everyone at Springer who made this volume possible. Special thanks go to Springer editor, Vaishali Damle, who initially suggested that the short course in Mexico City could serve as the basis for a Springer Brief.

Contents

Chapter 1
Basics

Abstract The study of inverse limits with set-valued functions was introduced in 2004 and has developed into a rich topic for research in topology. One path into this subject can be found by working through examples of such inverse limits. A natural starting point for such an undertaking is consideration of examples with a single bonding function on $[0, 1]$ having closed set values. In this chapter we include the basic definitions and theorems needed to read the remainder of the book much of which is driven by examples. We state and prove our theorems on $[0, 1]$; with minor modifications the proofs generally are valid in a much more general setting such as compact metric spaces or even compact Hausdorff spaces.

1.1 Introduction

In this chapter we introduce inverse limits with set-valued functions. Our aim is to provide a basis for the remainder of the book. Most of the material in this short book is devoted to inverse limits on the interval $[0, 1]$ with set-valued bonding functions. Such inverse limits are defined as subsets of the Hilbert cube, $Q = [0, 1]^\infty$. We assume that the reader has a working knowledge of compactness, connectedness, and continuity in metric spaces and some familiarity with the topology of product spaces. Such topics are likely to be covered in sufficient depth in an undergraduate course in topology or a senior-level course in analysis. The essentials of the background material may be found in most topology textbooks or in the appendix of the book *Inverse Limits: From Continua to Chaos* [2]. Most of the theorems are stated for inverse limits with upper semicontinuous bonding functions on closed subsets of $[0, 1]$ although, generally speaking, they hold in the setting of compact Hausdorff spaces and often the proofs given for $[0, 1]$ can easily be adapted to compact metric spaces or even compact Hausdorff spaces.

A metric space X is *compact* provided that if $\mathcal{G}$ is a collection of open sets covering X, then some finite subcollection of $\mathcal{G}$ covers X. A collection $\mathcal{A}$ of subsets of X is said to have the *finite intersection property* provided that if $\mathcal{B}$ is a finite

W.T. Ingram, *An Introduction to Inverse Limits with Set-valued Functions*,
SpringerBriefs in Mathematics, DOI 10.1007/978-1-4614-4487-9_1,
© W.T. Ingram 2012

subcollection of $\mathcal{A}$, then there is a point common to all the elements of $\mathcal{B}$. Our first theorem is a basic property of compact spaces that is likely to be well known to the reader, but we include its proof for the sake of completeness.

Theorem 1.1. *A metric space X is compact if and only if for each collection $\mathcal{A}$ of nonempty closed subsets of X having the finite intersection property there is a point common to all the elements of $\mathcal{A}$.*

Proof. Suppose X is compact and there is a collection $\mathcal{A}$ of nonempty closed subsets of X with the finite intersection property, but there is not a point common to all the elements of $\mathcal{A}$. Then, the collection $\mathcal{G} = \{X - A \mid A \in \mathcal{A}\}$ is a collection of open sets covering X. Some finite subcollection $\mathcal{H}$ of $\mathcal{G}$ covers X. Then, $\{X - h \mid h \in \mathcal{H}\}$ is a finite subcollection of $\mathcal{A}$ with no point in common, a contradiction.

On the other hand, suppose X is a metric space with the property that if $\mathcal{A}$ is a collection of closed subsets of X and $\mathcal{A}$ has the finite intersection property, then there is a point common to all the elements of $\mathcal{A}$. Suppose moreover that $\mathcal{G}$ is a collection of open sets covering X such that no finite subcollection of $\mathcal{G}$ covers X. Let $\mathcal{A} = \{X - g \mid g \in \mathcal{G}\}$; $\mathcal{A}$ is a collection of closed subsets of X. If $\mathcal{B}$ is a finite subcollection of $\mathcal{A}$, then there is point of X common to all the elements of $\mathcal{B}$, or else $\{X - B \mid B \in \mathcal{B}\}$ is a finite subcollection of $\mathcal{G}$ that covers X. Because $\mathcal{A}$ has the finite intersection property, by hypothesis there is an element of X common to all the elements of $\mathcal{A}$. Such a point belongs to no element of $\mathcal{G}$, a contradiction. □

Two subsets of a metric space are *mutually exclusive* provided they have no point in common; they are *mutually separated* provided they are mutually exclusive and neither contains a limit point of the other. A set is *connected* provided it is not the union of two mutually separated sets. By a *continuum*, we mean a compact-connected metric space; a compact set is a continuum if and only if it is not the union of two mutually exclusive compact sets.

1.2 Upper Semicontinuous Functions

If X is a metric space, 2^X denotes the collection of closed subsets of X, and $C(X)$ denotes the connected elements of 2^X. If each of X and Y is a metric space, a function f from X into 2^Y, denoted $f : X \to 2^Y$, is said to be *upper semicontinuous at the point $x \in X$* provided that if V is an open set in Y containing $f(x)$, then there is an open set U in X containing x such that if $t \in U$, then $f(t) \subseteq V$; f is called *upper semicontinuous* if it is upper semicontinuous at each point of X. If $f : X \to 2^Y$ is a function, let $G(f) = \{(x, y) \in X \times Y \mid y \in f(x)\}$; $G(f)$ is called the *graph* of f. If $f : X \to 2^Y$ is a function and $A \subseteq X$, $f(A)$ denotes $\{y \in Y \mid,$ there is a point $x \in A$ such that $y \in f(x)\}$; f is said to have a *surjective graph* or, simply, to be *surjective*, provided $f(X) = Y$. If f is an upper semicontinuous function having all of its values degenerate sets and

$f(x) = \{y\}$, we normally write $f(x) = y$; such upper semicontinuous functions are continuous, and we refer to continuous functions as *mappings*. We use Id_X to denote the *identity* on X, i.e., $Id_X(x) = x$ for each $x \in X$. When the domain should be clear from context, we normally shorten Id_X to Id. Our first theorem can be found in [1, Theorem 2.1, p. 120] as well as in [2, Theorem 105, p. 78] in a more general setting. We include it in the setting of compact metric spaces.

Theorem 1.2. *Suppose each of X and Y is a compact metric space and M is a subset of $X \times Y$ such that if $x \in X$, then there is a point $y \in Y$ such that $(x, y) \in M$. Then, M is closed if and only if there is an upper semicontinuous function $f : X \to 2^Y$ such that $M = G(f)$.*

Proof. We first show that if $f : X \to 2^Y$ is an upper semicontinuous function, then $G(f)$ is closed. Let $p = (p_1, p_2)$ be a point of $X \times Y$ that is not in $G(f)$. Then, $p_2 \notin f(p_1)$; so, because compact metric spaces are regular [2, Lemma 268, p. 179], there are mutually exclusive open sets V and W in Y such that $p_2 \in V$ and $f(p_1) \subseteq W$. Because f is upper semicontinuous, there is an open subset U of X containing p_1 such that if $t \in U$, then $f(t) \subseteq W$. Thus, $U \times V$ is an open subset of $X \times Y$ containing p that does not intersect $G(f)$. It follows that $G(f)$ is closed.

Assume that M is closed and, for each x in X, define $f(x)$ to be $\{y \in Y \mid (x, y) \in M\}$. Because $f(x)$ is the intersection of M with the closed set $\{x\} \times Y$, $f(x)$ is closed for each x in X. To see that f is upper semicontinuous, suppose x is in X and V is an open set in Y containing $f(x)$. If f is not upper semicontinuous at x, then for each open set U containing x, there exist points z of U and (z, y) of M such that y is not in V. For each open set U containing x, denote by M_U the nonempty set of all points (p, q) of M such that p is in $\overline{U}$ and q is not in V; M_U is closed. Observe that if U and U' are open sets containing x and $U \subseteq U'$, then $M_U \subseteq M_{U'}$. From this, it follows that the collection $\mathcal{M} = \{M_U \mid U \text{ is open in } X\}$ has the finite intersection property. Because $X \times Y$ is compact, by Theorem 1.1 there is a point (a, b) common to all the sets in $\mathcal{M}$. Because each element of $\mathcal{M}$ is a subset of M, (a, b) belongs to M so $b \in f(a)$. Because x is the only point common to all the sets $\overline{U}$, $a = x$. However, b is not in V, contradicting the fact that b belongs to $f(x)$.   □

There are a couple of additional properties of upper semicontinuous functions that we use in this book. We include them here to help familiarize the reader with such functions. We leave the proof of Theorem 1.3 to the reader.

Theorem 1.3. *If each of X and Y is a compact metric space and $f : X \to 2^Y$ is upper semicontinuous, then $\varphi : X \to 2^{X \times Y}$ given by $\varphi(x) = \{x\} \times f(x)$ and $\psi : X \to 2^{Y \times X}$ given by $\psi(x) = f(x) \times \{x\}$ are upper semicontinuous.*

Suppose X, Y, and Z are metric spaces and $f : X \to 2^Y$ and $g : Y \to 2^Z$ are set-valued functions. By the *composition* of g with f, denoted $g \circ f$, we mean that the function from X into 2^Z given by $g \circ f(x) = \{z \in Z \mid \text{there is a point } y \in Y \text{ such that } y \in f(x) \text{ and } z \in g(y)\}$. Sometimes, we shorten $g \circ f$ to gf.

Theorem 1.4. *Suppose X, Y, and Z are metric spaces. If $f : X \to 2^Y$ and $g : Y \to 2^Z$ are upper semicontinuous, then $g \circ f$ is upper semicontinuous.*

Proof. Let x be a point of X and suppose V is an open set in Z that contains $g \circ f(x)$. Using that g is upper semicontinuous, for each $y \in f(x)$, choose an open subset W_y of Y containing y such that if $s \in W_y$, then $g(s) \in V$. Let $W = \bigcup_{y \in f(x)} W_y$. Then W is an open set and $f(x) \subseteq W$. Because f is upper semicontinuous, there is an open set U containing x such that if $t \in U$, then $f(t) \subseteq W$. Suppose $t \in U$ and p is a point of $f(t)$. Then, there is a point $y \in f(x)$ such that $p \in W_y$, so $g(p) \in V$. Therefore, $g(f(t)) \subseteq V$.   □

1.3   Inverse Limits

A *sequence* is a function defined on a set of nonnegative integers, normally the set of all positive integers. We generally denote sequences in boldface type and the terms of sequences in italics. To simplify our notation, we adopt the following convention. If a is a sequence, we denote $a(i)$ by a_i. If a is a sequence, we sometimes denote a by listing its terms $a_1, a_2, a_3, \ldots$. If X is a sequence such that $X_i = [0, 1]$ for each positive integer i, we sometimes denote $\prod_{i=1}^n X_i$ by $[0, 1]^n$; the product, $\prod_{i>0} X_i$, of the sequence is the *Hilbert cube*, Q. The points of Q are sequences of numbers from $[0, 1]$; a metric d for Q that is consistent with the product topology on Q is given by $d(x, y) = \sum_{i>0} |x_i - y_i|/2^i$. Additional information on Q may be found in the appendix of [2]. In the case that X is a sequence of closed subsets of $[0, 1]$, the points of $\prod_{i>0} X_i$ are sequences x such that $x_i \in X_i$ for each positive integer i, i.e., $\prod_{i>0} X_i$ is a subset of Q, and, as such, it inherits a topology from Q. Although a point x of Q is a sequence, we normally enclose its terms in parenthesis to indicate that it is a point of a product space, i.e., $x = (x_1, x_2, x_3, \ldots)$. If X is a sequence of closed subsets of $[0, 1]$ and f is a sequence of functions such that $f_i : X_{i+1} \to 2^{X_i}$ for each positive integer i, the pair of sequences $\{X, f\}$ is called an *inverse limit sequence*; the sets $X_1, X_2, X_3, \ldots$ are called *factor spaces*; the functions $f_1, f_2, f_3, \ldots$ are called *bonding functions*. In the case that the factor spaces are understood (e.g., all the factor spaces are $[0, 1]$ as is the case throughout most of this book), we may refer to f as an inverse limit sequence. By the *inverse limit* of the inverse limit sequence $\{X, f\}$, we mean $\{x \in Q \mid x_j \in f_j(x_{j+1})$ for each positive integer $j\}$. If $\{X, f\}$ is an inverse limit sequence, we denote its inverse limit by $\varprojlim\{X, f\}$, although normally we shorten this to $\varprojlim f$. In the case that the sequences X and f are constant sequences, i.e., there are a closed subset X of $[0, 1]$ and a function $f : X \to 2^X$ such that $X_i = X$ and $f_i = f$ for each positive integer i, we still denote the inverse limit by $\varprojlim f$. These are called *inverse limits with only one bonding function*. Most of the inverse limits we discuss in this book, particularly the examples, are inverse limits with only one bonding function.

Throughout, we use $\mathbb{N}$ to denote the set of positive integers. If X is a sequence of closed subsets of $[0, 1]$ and $A \subseteq \mathbb{N}$, denote by $p_A : \prod_{i>0} X_i \to \prod_{i \in A} X_i$ the function given by $p_A(x) = y$ where $y_i = x_i$ for each $i \in A$. For any $A \subseteq \mathbb{N}$, p_A is a mapping called a *projection*. In the case that $A = \{n\}$, we normally denote p_A by p_n. For inverse limits, we adopt the convention of using π_A to denote the restriction of p_A to the inverse limit.

1.4   Some Basic Examples

In this section we provide some elementary examples to illustrate the definition of an inverse limit as well as to indicate some of the inverse limits that may be obtained using only one simple bonding function on the interval $[0, 1]$. Most of the functions in our examples have graphs that are easy to draw; however, for easy reference, we provide a picture showing the graph in many of our examples. Later in the book, as the inverse limits become more complicated, we sometimes provide a model for the inverse limit. Each of the examples in this section is nonempty. This follows from Theorem 1.6 from Sect. 1.5; however, even without that theorem, it can be seen that each inverse limit in this section contains the point $(0, 0, 0, \dots)$ because in each case, $0 \in f(0)$.

Example 1.1 (The Hilbert cube $\mathcal{Q}$). Let $f : [0, 1] \to 2^{[0,1]}$ be given by $f(t) = [0, 1]$ for each $t, 0 \le t \le 1$. Then $\varprojlim f = \mathcal{Q}$.

The Cantor set is another familiar compactum that is obtainable as an inverse limit on $[0, 1]$ using a single set-valued bonding function.

Example 1.2 (The Cantor set). Let $f : [0, 1] \to 2^{[0,1]}$ be given by $f(t) = \{0, 1\}$ for $0 \le t \le 1$. Then $\varprojlim f$ is a Cantor set. (See Fig. 1.1 for the graph of f.)

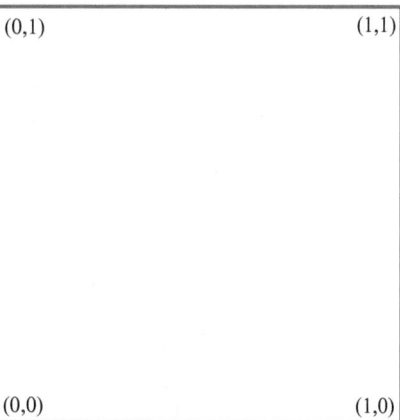

(0,1)   (1,1)

(0,0)   (1,0)

Fig. 1.1 The graph of the bonding function in Example 1.2. The inverse limit is a Cantor set

Fig. 1.2 The bonding
mapping in Example 1.4. The
inverse limit is degenerate

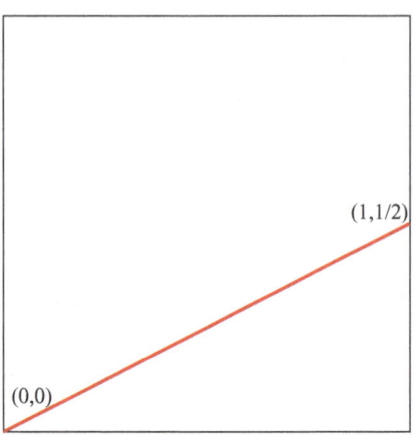

(1,1/2)

(0,0)

Proof. $M = \varprojlim f$ cannot contain a point with first coordinate strictly between 0 and 1; thus, $M = \prod_{i>0} X_i$ where $X_i = \{0, 1\}$ for each positive integer i.   □

The reader should note that we obtain the same inverse limit if we restrict the bonding function in Example 1.2 to $\{0, 1\}$.

Example 1.3. Let $X = \{0, 1\}$ and $f : X \to 2^X$ be given by $f(x) = X$ for each $x \in X$. Then, $\varprojlim f$ is a Cantor set.

Even using a mapping as a bonding function, the inverse limit may be *degenerate*, i.e., a single point, as the next example shows.

Example 1.4 (A degenerate inverse limit). Let $f : [0, 1] \to [0, 1]$ be given by $f(t) = t/2$. Then, $\varprojlim f = \{(0, 0, 0, \dots)\}$. (See Fig. 1.2 for the graph of f.)

Proof. Observe that if $0 < t \leq 1/2$, then there is a positive integer n such that $f^{-n}(t) > 1/2$. If $z > 1/2$ and $t \in [0, 1]$, then $f(t) \neq z$. It follows that no point of the inverse limit can have a positive coordinate.   □

Example 1.5 (A two-point inverse limit). Let $f : [0, 1] \to 2^{[0,1]}$ be given by $f(t) = t/2$ for $t < 1/2$, $f(1/2) = \{1/4, 3/4\}$ and $f(t) = 1/2 + t/2$ for $t > 1/2$. Then, $\varprojlim f = \{(0, 0, 0, \dots), (1, 1, 1, \dots)\}$. (See Fig. 1.3 for the graph of f.)

Proof. Let $M = \varprojlim f$. Let $f_1 : [0, 1/2] \to [0, 1/2]$ be given by $f_1(t) = t/2$ and $f_2 : [1/2, 1] \to [1/2, 1]$ be given by $f_2(t) = 1/2 + t/2$. Let $M_1 = \varprojlim f_1$ and $M_2 = \varprojlim f_2$. It is easy to see that $M_1 \cup M_2 \subseteq M$. If $x \in M$ and $x \notin M_1$, then there is a positive integer i such that $x_i \notin [0, 1/2]$. It follows that $x_j \in [1/2, 1]$ for each positive integer j and consequently $x \in M_2$. Thus, $M = M_1 \cup M_2$. Clearly, no point of M has first coordinate between $1/4$ and $3/4$. The remainder of the proof that M_1 and M_2 are degenerate follows in much the same way as Example 1.4 was shown to be degenerate.   □

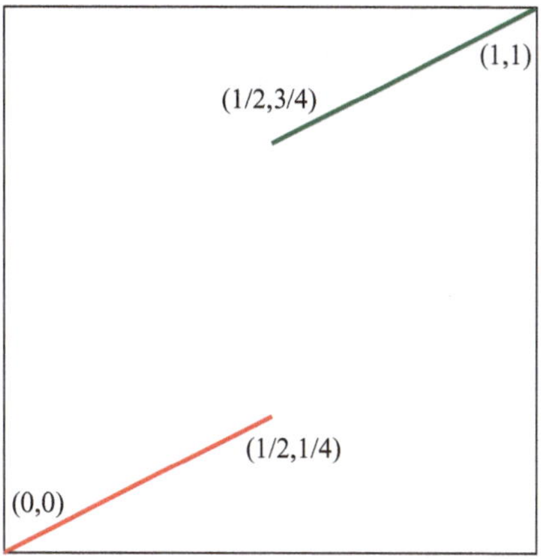

Fig. 1.3 The graph of the bonding function in Example 1.5. The inverse limit consists of two points

Fig. 1.4 The graph of the bonding function in Example 1.6. The inverse limit is a set containing only three points

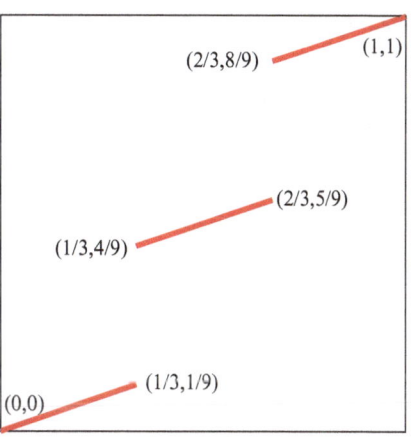

In the case that f is a sequence of surjective mappings, the inverse limit is never degenerate. In fact, it is uncountable. This follows from Theorem 1.7 in the next section.

The function in our next example has an inverse limit that is a point set containing only three points. We provide it without proof because the proof is similar to that of Example 1.5. It should be clear to the reader how to obtain any finite set as an inverse limit on $[0, 1]$ with a single set-valued function.

Example 1.6 (A three-point inverse limit). Let $f : [0, 1] \rightarrow 2^{[0,1]}$ be given by $f(t) = t/3$ for $0 \leq t < 1/3, f(1/3) = \{1/9, 4/9\}, f(t) = t/3 + 1/3$ for $1/3 < t < 2/3, f(2/3) = \{5/9, 8/9\}$, and $f(t) = t/3 + 2/3$ for $2/3 < t \leq 1$. Then $\varprojlim f = \{(0, 0, 0, \dots), (1/2, 1/2, 1/2, \dots), (1, 1, 1, \dots)\}$. (See Fig. 1.4 for the graph of f.)

In our next example, the inverse limit is a convergent sequence.

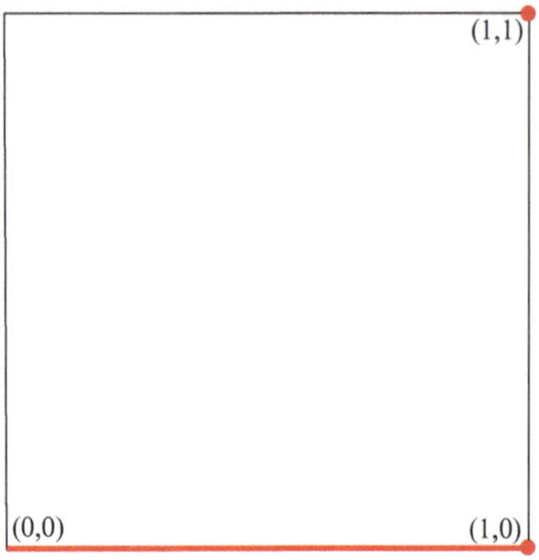

Fig. 1.5 The graph of the
bonding function in
Example 1.7. The inverse
limit is a simple convergent
sequence

Fig. 1.6 The graph of the
bonding function in
Example 1.8. The inverse
limit is connected even
though the graph of the
bonding function is not
connected

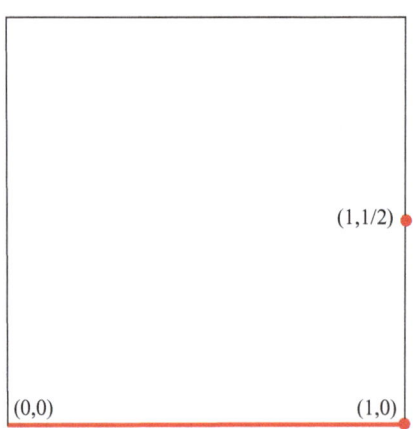

Example 1.7 (A convergent sequence). Let $f : [0, 1] \rightarrow 2^{[0,1]}$ be given by $f(t) = 0$ for $0 \leq t < 1$ and $f(1) = \{0, 1\}$. Then $\varprojlim f = \{(0, 0, 0, \dots), (1, 1, 1, \dots),$ $(0, 1, 1, 1, \dots), (0, 0, 1, 1, \dots), \dots\}$. Note that if $g : \{0, 1\} \rightarrow 2^{\{0,1\}}$ is given by $g(0) = 0$ and $g(1) = \{0, 1\}$, $\varprojlim f = \varprojlim g$. (See Fig. 1.5 for the graph of f.)

In Example 1.7, the fact that 1 is a fixed point for f, i.e., $1 \in f(1)$, plays a crucial role as we see from the following example where the bonding function is similar but the inverse limit is nowhere near as rich. Note that the inverse limit from Example 1.8 is connected even though the graph of the bonding function is not connected. In Theorem 2.2 we show that this cannot happen if the bonding functions are surjective.

Example 1.8 (A connected inverse limit from bonding function with a nonconnected graph). Let $f : [0, 1] \rightarrow 2^{[0,1]}$ be given by $f(t) = 0$ for $0 \leq t < 1$ and $f(1) = \{0, 1/2\}$. Then $\varprojlim f = \{(0, 0, 0, \dots)\}$. (See Fig. 1.6 for the graph of f.)

Proof. Suppose $x \in Q$ and $x_1 > 0$. If $x_1 \neq 1/2$, then, because $x_1 \notin f(t)$ for any $t \in [0, 1]$, $x \notin \varprojlim f$. If $x_1 = 1/2$, then $x_1 \in f(t)$ only for $t = 1$. But $1 \notin f(s)$ for any $s \in [0, 1]$, so once again, $x \notin \varprojlim f$. It follows that $(0, 0, 0, \dots)$ is the only point of $\varprojlim f$. □

1.5 A Fundamental Existence Theorem

Because our emphasis in this book is on inverse limits on $[0, 1]$ with upper semicontinuous bonding functions, we state and prove our theorems for inverse limit sequences on closed subsets of $[0, 1]$. Most of these theorems carry over to inverse limits with upper semicontinuous functions on compact Hausdorff spaces, and for the most part, their proofs require only minor modification. The corresponding theorems in more generality may be found in [2]. Suppose $\{X, f\}$ is an inverse limit sequence such that X_i is a closed subset of $[0, 1]$ for each positive integer i. We adopt the following notation and use it extensively in this book. If n is a positive integer, let $G_n = \{x \in Q \mid x_i \in f_i(x_{i+1}) \text{ for } 1 \leq i \leq n\}$.

Theorem 1.5. *If X is a sequence of closed subsets of $[0, 1]$ and f is a sequence of upper semicontinuous functions such that $f_i : X_{i+1} \to 2^{X_i}$, then G_n is a nonempty compact subset of $\prod_{i>0} X_i$.*

Proof. If $p \in X_{n+1}$, there is a point $x_n \in X_n$ such that $x_n \in f_n(p)$. Continuing inductively, there is a finite set $x_1, x_2, \dots, x_n$ such that $x_i \in X_i$ and $x_i \in f_i(x_{i+1})$ for $1 \leq i < n$. Choose $y \in \prod_{i>0} X_i$. The point $(x_1, x_2, \dots, x_n, p, y_{n+2}, y_{n+3}, \dots)$ is a point of G_n, so G_n is nonempty.

To show that G_n is compact, we only need to show that G_n is a closed subset of the compact metric space Q. Suppose $x \in Q - G_n$. Then, there is a positive integer $k \leq n$ such that $x_k \notin f_k(x_{k+1})$. There exist mutually exclusive open sets O and U such that $x_k \in O$ and $f_k(x_{k+1}) \subseteq U$. Because f_k is upper semicontinuous, there is an open set V containing x_{k+1} such that if $t \in V$, then $f(t) \subseteq U$. Then, because the projection p_i is continuous for each i, $p_k^{-1}(O) \cap p_{k+1}^{-1}(V)$ is an open set containing x that contains no point of G_n. It follows that G_n is closed. □

An alternate proof of Theorem 1.5 can be constructed employing Theorem 1.2 and the observation made by Gerardo Acosta that $G_n = p_{\{1,2\}}^{-1}(G(f_1^{-1})) \cap p_{\{2,3\}}^{-1}(G(f_2^{-1})) \cap \dots \cap p_{\{n,n+1\}}^{-1}(G(f_n^{-1}))$.

Theorem 1.6. *If X is a sequence of closed subsets of $[0, 1]$ and f is a sequence of upper semicontinuous functions such that $f_i : X_{i+1} \to 2^{X_i}$, then $\varprojlim f$ is a nonempty compact metric space.*

Proof. For each positive integer n, $G_{n+1} \subseteq G_n$ and, by Theorem 1.5, G_n is compact. Thus, $G_1, G_2, G_3, \dots$ is a nested sequence of nonempty compact subsets of Q. By Theorem 1.1, there is a point of Q belonging to $\bigcap_{i>0} G_i$, a compact set. But $\varprojlim f = \bigcap_{i>0} G_i$. □

In the case that the bonding functions are surjective, we do not need Theorem 1.6 to see that the inverse limit is nonempty. It is a consequence of the following theorem that does not require upper semicontinuity of the bonding functions.

Theorem 1.7. *Suppose X is a sequence of closed subsets of $[0, 1]$ and f is a sequence of surjective functions such that $f_i : X_{i+1} \rightarrow 2^{X_i}$ for each i. If n is a positive integer and $p \in X_n$, then there is a point $x \in \varprojlim f$ such that $x_n = p$.*

Proof. We construct a point x as follows. Let $x_n = p$. There is a point x_{n-1} of X_{n-1} such that $x_{n-1} \in f_n(p)$. Similarly, it follows that there is a finite sequence $x_1, x_2, x_3, \ldots, x_{n-1}$ such that $x_i \in f_i(x_{i+1})$ for $1 \leq i < n-1$. Because f_n is surjective, there is a point $x_{n+1} \in X_{n+1}$ such that $p \in f_n(x_{n+1})$. Continuing by induction and employing the surjectivity of the bonding functions, we construct a sequence $x_{n+1}, x_{n+2}, x_{n+3}, \ldots$ such that $x_j \in f_j(x_{j+1})$ for $j > n$. Thus, we obtain a point x of $\varprojlim f$ such that $x_n = p$. $\square$

A stronger theorem holds as we see in Theorem 1.8. We leave its proof to the reader. Before stating the theorem, we introduce some notation. If $\{X, f\}$ is an inverse limit sequence and m and n are positive integers with $m < n$, by f_{mn}, we mean $f_m \circ f_{m+1} \circ \cdots \circ f_{n-1}$. Note that $f_{mn} : X_n \rightarrow 2^{X_m}$. This notation is normally extended to f_{mn} for $m \leq n$ by letting f_{nn} denote the identity on X_n.

Theorem 1.8. *Suppose X is a sequence of closed subsets of $[0, 1]$ and f is a sequence of surjective upper semicontinuous functions such that $f_i : X_{i+1} \rightarrow 2^{X_i}$ for each i. If $m, n \in \mathbb{N}$ with $m < n$ and $q \in X_m$ and $p \in X_n$ with $q \in f_{mn}(p)$, there is a point $x \in \varprojlim f$ such that $x_m = q$ and $x_n = p$.*

1.6   Some Elementary Basic Theorems

The following theorem is easy to prove, and it will be used often throughout without a specific reference to it. In fact, we made use of it in Example 1.5.

Theorem 1.9. *Suppose C is a closed subset of $[0, 1]$. If $f : [0, 1] \rightarrow 2^{[0,1]}$ and $g : C \rightarrow 2^C$ are upper semicontinuous functions such that $G(g) \subseteq G(f)$, then $\varprojlim g \subseteq \varprojlim f$.*

A function $f : X \rightarrow Y$ is said to be 1–1 provided $f(x) \neq f(y)$ whenever $x \neq y$. A *homeomorphism* is mapping that is 1–1 with a continuous inverse. A 1–1 mapping of a compact subset of Q is a homeomorphism [2, Theorem 259, p. 178]. By an *arc* we mean a continuum that is homeomorphic to the interval $[0, 1]$.

Example 1.9 (Union of an arc and a convergent sequence). Let $f : [0, 1] \rightarrow 2^{[0,1]}$ be given by $f(t) = t$ for $0 \leq t < 1$ and $f(1) = \{0, 1\}$ and $g : \{0, 1\} \rightarrow 2^{\{0,1\}}$ be given by $g(0) = 0$ and $g(1) = \{0, 1\}$, as in Example 1.7. Then $\varprojlim f = \varprojlim Id \cup \varprojlim g$ and thus is the union of an arc and a convergent sequence. (See Fig. 1.7 for the graph of f.)

Fig. 1.7 The graph of the bonding function in Example 1.9. The inverse limit is the union of an arc and a simple convergent sequence

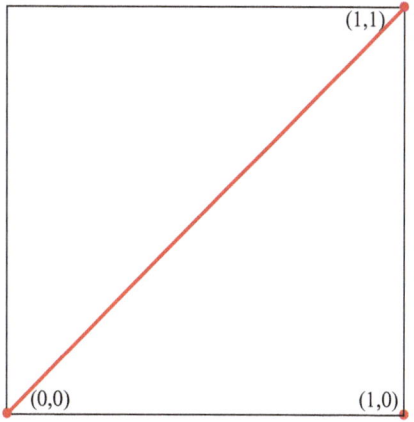

Proof. It is easy to see that $A = \varprojlim \boldsymbol{Id}$ is an arc; in fact, $h : [0,1] \to A$ given by $h(t) = (t, t, t, \ldots)$ is a homeomorphism. For $B = \varprojlim \boldsymbol{g}$, $B = \{(0,0,0,\ldots),$ $(1,1,1,\ldots), (0,1,1,1,\ldots), (0,0,1,1,1,\ldots), \ldots\}$. Because $G(Id) \cup G(g) \subseteq G(f)$, $\varprojlim \boldsymbol{Id} \cup \varprojlim \boldsymbol{g} \subseteq \varprojlim \boldsymbol{f}$. On the other hand, if $\boldsymbol{x} \in \varprojlim \boldsymbol{f}$ and $\boldsymbol{x} \notin \varprojlim \boldsymbol{Id}$, then $x_i \notin (0,1)$ for any positive integer i. Thus, if $\boldsymbol{x} \neq (0,0,0,\ldots)$, then there is a positive integer j such that $x_i = 0$ for $1 \leq i \leq j$ and $x_i = 1$ for $i > j$. Such a point is in B. $\square$

Theorem 1.10. *Suppose X is a sequence of closed subsets of $[0,1]$ and $\boldsymbol{f}$ is a sequence of upper semicontinuous functions such that $f_i : X_{i+1} \to 2^{X_i}$ for each positive integer i. If $\boldsymbol{x} \in \varprojlim \boldsymbol{f}$ and m and n are positive integers with $m < n$, then $x_m \in f_{mn}(x_n)$ (or, stated in terms of projections, $\pi_m(\boldsymbol{x}) \in f_{mn}(\pi_n(\boldsymbol{x})))$.*

We revisit the interaction between the projections and set-valued bonding functions in Sect. 3.3.

References

1. Ingram, W.T., Mahavier, W.S.: Inverse limits of upper semi-continuous set valued functions. Houston J. Math. **32**, 119–130 (2006)
2. Ingram, W.T., Mahavier, W.S.: Inverse limits: From continua to Chaos. In: Developments in Mathematics, vol. 25. Springer, New York (2012)

Chapter 2
Connectedness

Abstract A fundamental question about inverse limits with set-valued bonding functions relates to the connectedness of the inverse limit. For inverse limits on compact, connected factor spaces with bonding functions that are mappings, the inverse limit is always connected. However, for inverse limits with set-valued functions as bonding functions, the inverse limit is rarely connected. One might suspect that this is due to the fact that the graph of an upper semicontinuous function on a compact, connected space can fail to be connected, but the reasons go much deeper. In this chapter we study connectedness of inverse limits on [0, 1] with set-valued functions.

2.1 Introduction

All but one of the examples from Chap. 1 were not connected, and after examining those examples, one might conjecture that if the graphs of the bonding functions are connected, then the inverse limit is connected. This is not the case, and in Example 2.1 of Sect. 2.2, we provide the first of several examples of functions having connected graphs and a nonconnected inverse limit. In fact, there are examples of inverse limit sequences with only one bonding function such that the inverse limit is totally disconnected even though the graph of the bonding function is connected. Although their example is beyond the scope of this book, it has been shown by Sina Greenwood and Judy Kennedy that there exists an inverse limit on [0, 1] with a single surjective set-valued bonding function having a connected graph such that the inverse limit is a Cantor set. They go on to show that in some sense most inverse limits with upper semicontinuous bonding functions are not connected. In this chapter we discuss connectedness of the inverse limit with upper semicontinuous bonding functions. Much of our discussion relates, in one way or another, the unsolved problem of characterizing connectedness of inverse limits with upper semicontinuous bonding functions in terms of the bonding functions.

W.T. Ingram, *An Introduction to Inverse Limits with Set-valued Functions,*
SpringerBriefs in Mathematics, DOI 10.1007/978-1-4614-4487-9_2,
© W.T. Ingram 2012

One of the major problems in the theory of inverse limits with set-valued functions is the question of under what conditions is the inverse limit connected (see Problem 6.1 in Chap. 6). A word about a solution to the problem of characterizing connectedness is in order. In the next section, we actually present a solution to the problem. However, the solution is not very satisfying because it is not given specifically in terms of the nature of the bonding functions. Ideally, a solution would allow us to determine the connectedness of the inverse limit by an examination of the bonding functions. From this perspective, the problem remains unsolved even in the case that each factor space is the interval $[0, 1]$.

2.2 A Characterization of Connectedness

We begin our discussion of connectedness with a theorem that characterizes this property for inverse limits. Unfortunately, when the bonding functions are set-valued, it is rarely easy to verify that the hypothesis of Theorem 2.1 is satisfied, so the theorem is not very useful except under special circumstances. As with many theorems in this book, the following theorem holds in a much more general setting than we state. For a more general theorem, see [6, Theorem 116, p. 85] where it is shown that the connectedness of the inverse limit follows from the connectedness of the terms of the sequence G. Recall from Chap. 1 that, for a sequence X of closed subsets of $[0, 1]$ and a sequence f of upper semicontinuous functions such that $f_i : X_{i+1} \to 2^{X_i}$ for each positive integer i, G_n is defined to be $\{x \in \mathcal{Q} \mid x_i \in f_i(x_{i+1}) \text{ for } 1 \leq i \leq n\}$.

Theorem 2.1. *Suppose X is a sequence of closed subsets of $[0, 1]$ and f is a sequence of upper semicontinuous functions such that $f_i : X_{i+1} \to 2^{X_i}$ for each positive integer i. Then, $\varprojlim f$ is a continuum if and only if G_n is connected for each positive integer n.*

Proof. Let $M = \varprojlim f$. We showed in Theorem 1.5 that G_n is compact for each positive integer n and $M = \bigcap_{n>0} G_n$. Thus, if G_n is connected for each $n \in \mathbb{N}$, then M is a continuum, being the intersection of a nested sequence of subcontinua of $\mathcal{Q}$.

On the other hand, if M is connected, then $\pi_{\{1,2,\dots,n+1\}}(M)$ is connected for each positive integer n because $\pi_{\{1,2,\dots,n+1\}}$ is a mapping. However, $G_n = \pi_{\{1,2,\dots,n+1\}}(M) \times \prod_{i>n+1} X_i$, a connected set because it is a product of two connected sets. $\qquad\square$

The proof of Theorem 2.1 makes use of the continuity of π_A where $A \subseteq \mathbb{N}$ to conclude that the image of a connected set under π_A is connected. This leads to our next theorem that was first observed by Van Nall in [10]. One can often obtain information about inverse limits with mappings by examining composites of the bonding maps, especially if the inverse limit in question is produced by a single bonding map. As we shall see later, this is rarely the case when the bonding

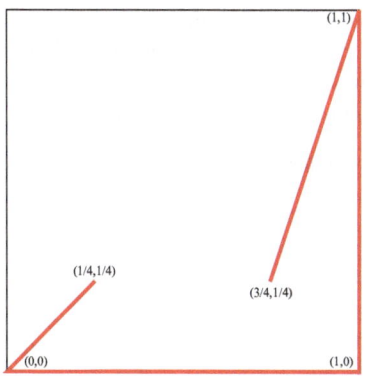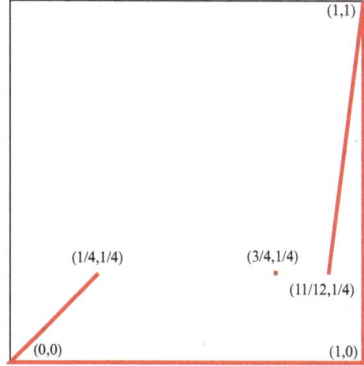

Fig. 2.1 Graph of the bonding function f (*left*) in Example 2.1 and f^2 (*right*)

functions are set-valued. However, Theorem 2.2 provides one of the few cases in the theory of inverse limits where composites provide some insight into the nature of the inverse limit when the bonding functions are set-valued.

Theorem 2.2 (Nall). *Suppose X is a sequence of closed subsets of $[0,1]$ and f is a sequence of surjective set-valued functions such that $f_i : X_{i+1} \to 2^{X_i}$ is upper semicontinuous for each positive integer i. If m and n are positive integers such that $m < n$ and $G(f_{mn})$ is not connected, then $\varprojlim f$ is not connected.*

Proof. If $\varprojlim f$ is connected and $m, n \in \mathbb{N}$ with $m < n$, then $\pi_{\{m,n\}}(\varprojlim f)$ is connected. However, by using Theorem 1.8, we see that $\pi_{\{m,n\}}(\varprojlim f) = (G(f_{mn}))^{-1}$, so $G(f_{mn})$ is connected, a contradiction. $\qquad\square$

One consequence of Theorem 2.2 is that if an inverse limit with surjective bonding functions is connected, then each of the bonding functions (including compositions) has a connected graph. In Example 1.8, we saw that an inverse limit using nonsurjective bonding functions with graphs that are not connected can be connected. Our first example of this chapter is a surjective upper semicontinuous function having a connected graph and an inverse limit that is not connected. It was first published in [5] but with a different proof that it is not connected.

Example 2.1. Let $f : [0,1] \to 2^{[0,1]}$ be given by $f(t) = \{0, t\}$ for $0 \le t \le 1/4$, $f(t) = 0$ for $1/4 < t < 3/4$, $f(t) = \{3t - 2, 0\}$ for $3/4 \le t < 1$, and $f(1) = [0,1]$. Then, $G(f)$ is connected, but $\varprojlim f$ is not connected. (See the graph on the left in Fig. 2.1 for $G(f)$.)

Proof. It is not difficult to show that $f^2(t) = \{t, 0\}$ for $0 \le t \le 1/4$, $f^2(t) = 0$ for $1/4 < t < 3/4$, $f^2(3/4) = \{1/4, 0\}$, $f^2(t) = 0$ for $3/4 < t < 11/12$, and $f^2(t) = \{9t - 8, 0\}$ for $11/12 \le t \le 1$ ($G(f^2)$ are pictured on the right in Fig. 2.1). Because $(3/4, 1/4)$ is an isolated point of $G(f^2)$, it follows from Theorem 2.2 that $\varprojlim f$ is not connected. $\qquad\square$

To conclude this section, we present a second simple theorem, this one characterizing connectedness of an inverse limit on intervals with a single bonding function. Nall included this theorem for inverse limits on Hausdorff continua in [10, Theorem 3.3, p. 171]. The proof given here is based on his proof.

Theorem 2.3. *Suppose $f : [0, 1] \to 2^{[0,1]}$ is an upper semicontinuous function that is surjective. Then, $\varprojlim f$ is connected if and only if $\varprojlim f^{-1}$ is connected.*

Proof. Suppose $n \in \mathbb{N}$. Observe that $h : \mathcal{Q} \to \mathcal{Q}$ given by $h(x) = (x_{n+1}, x_n, \ldots, x_1, x_{n+2}, x_{n+3}, \ldots)$ is a homeomorphism. Let $g = f^{-1}$. If $G_n = \{x \in \mathcal{Q} \mid x_i \in f_i(x_{i+1})$ for $1 \le i \le n\}$ and $H_n = \{x \in \mathcal{Q} \mid x_i \in g(x_{i+1})$ for $1 \le i \le n\}$, then $h(G_n) = H_n$, and thus H_n is connected if and only if G_n is connected. The theorem now follows from Theorem 2.1.   □

In Theorem 2.10 below, we show that a theorem similar to Theorem 2.3 holds if the constant sequence f is replaced by a sequence of mappings. It would be interesting to know if the constant sequence in Theorem 2.3 can be replaced by a sequence of surjective upper semicontinuous functions (see Problem 6.7).

2.3 A Basic Connectedness Theorem

Theorem 2.7 as stated below for sequences of upper semicontinuous functions on subintervals of $[0, 1]$ appeared in Mahavier's original paper [7] on inverse limits with subsets of $[0, 1] \times [0, 1]$. It was generalized in [5] to inverse limits of sequences of compact, connected Hausdorff spaces with upper semicontinuous bonding functions and in [6] to consistent inverse limit systems on compact, connected Hausdorff spaces with upper semicontinuous bonding functions over directed sets. Although we state and prove it for inverse limits on subintervals of $[0, 1]$, we give a different proof of the theorem based on the following theorems not found in those references.

Suppose that each of X and Y is a continuum. A mapping $f : X \to Y$ is said to be *monotone* provided $f^{-1}(y)$ is connected for each $y \in f(X)$. In the case that f^{-1} is a surjective mapping, Theorem 2.4 follows from the well-known theorem that a surjective map of continua is monotone if and only if the preimage of each subcontinuum of the range is a subcontinuum of the domain [9, Exercise 8.46, p. 137]. We present a proof of a slightly different theorem based on a direct use of upper semicontinuity. Recall our notation that if $f : X \to 2^Y$ and $A \subseteq X$, by $f(A)$ we mean $\{y \in Y \mid$, there is a point $x \in A$ such that $y \in f(x)\}$.

Theorem 2.4. *Suppose X is a continuum, Y is a compact metric space, and $f : X \to C(Y)$ is an upper semicontinuous function. Then, $f(X)$ is a continuum.*

Proof. Because $G(f)$ is compact and $f(X) = p(G(f))$ where p is the projection of $X \times Y$ onto Y, $f(X)$ is compact. If $f(X)$ is not connected, then there are two mutually exclusive compact sets H and K such that $f(X) = H \cup K$. The normality of metric spaces provides two mutually exclusive open sets U and V

such that $H \subseteq U$ and $K \subseteq V$. If $x \in X$, $f(x)$ is connected so $f(x) \subseteq H$ or $f(x) \subseteq K$. Let $X_H = \{x \in X \mid f(x) \subseteq H\}$ and $X_K = \{x \in X \mid f(x) \subseteq K\}$. Then, $X = X_H \cup X_K$, and no point of X belongs to both X_H and X_K. If $t \in X_H$, then $f(t) \subseteq U$. Because f is upper semicontinuous at t, there is an open set W containing t such that if $s \in W$, then $f(s) \subseteq U$. Then, $W \subseteq X_H$, and it follows that X_H is open. Therefore, X_K is closed. A similar argument yields that X_H is closed. This involves a contradiction because the continuum X is not the union of two mutually exclusive closed sets.   □

Theorem 2.5. *Suppose X is a continuum and Y is a compact metric space. If $f : X \to C(Y)$ is upper semicontinuous, then $G(f)$ is a continuum.*

Proof. Let $\varphi : X \to X \times Y$ be function given by $\varphi(x) = \{x\} \times f(x)$; φ is upper semicontinuous by Theorem 1.3. Then $G(f) = \varphi(X)$ and $\varphi(X)$ is a continuum by Theorem 2.4.   □

Next we extend our notion of the graph of an upper semicontinuous function in the following way. Suppose $\{X_1, X_2, \ldots, X_{n+1}\}$ is a finite collection of metric spaces and $\{f_1, f_2, \ldots, f_n\}$ is a finite collection of functions such that $f_i : X_{i+1} \to 2^{X_i}$ for $1 \leq i \leq n$. Define $G'(f_1, f_2, \ldots, f_n) = \{x \in \prod_{i=1}^{n+1} X_i \mid x_i \in f_i(x_{i+1})$ for $1 \leq i \leq n\}$. Note that if $f : X \to 2^Y$ is a function, $G'(f) = (G(f))^{-1} = G(f^{-1})$.

Lemma 2.1. *Suppose $\{X_1, X_2, \ldots, X_{n+1}\}$ is a finite collection of closed subsets of $[0, 1]$ and $\{f_1, f_2, \ldots, f_n\}$ is a finite collection of upper semicontinuous functions such that $f_i : X_{i+1} \to 2^{X_i}$ for $1 \leq i \leq n$. Then, $G'(f_1, f_2, \ldots, f_n)$ is compact.*

Proof. For $i > n + 1$, let $X_i = X_{n+1}$ and $f_i = Id_{X_{n+1}}$. By Theorem 1.5, $G_n = \{x \in \prod_{i>0} X_i \mid x_i \in f_i(x_{i+1})\}$ is nonempty and compact. Because $G'(f_1, f_2, \ldots, f_n) = \pi_{\{1,2,\ldots,n+1\}}(G_n)$, the conclusion follows.   □

Lemma 2.2. *Suppose $\{X_1, X_2, \ldots, X_{n+1}\}$ is a finite collection of continua, $\{f_1, f_2, \ldots, f_n\}$ is a finite collection of upper semicontinuous functions such that $f_i : X_{i+1} \to 2^{X_i}$ for $2 \leq i \leq n$, and $f_1 : X_2 \to C(X_1)$. If $G'(f_2, f_3, \ldots, f_n)$ is connected, then $G'(f_1, f_2, \ldots, f_n)$ is a continuum.*

Proof. In light of Lemma 2.1, we only need to show that $G'(f_1, f_2, \ldots, f_n)$ is connected. Let p denote the projection of $\prod_{i=2}^{n+1} X_i$ to its first factor space X_2. Because p is a mapping, $f_1 \circ p$ is upper semicontinuous. The function $\psi : G'(f_2, f_3, \ldots, f_n) \to G'(f_1, f_2, \ldots, f_n)$ given by $\psi(x) = f_1(p(x)) \times \{x\}$ is upper semicontinuous. It follows that $G'(f_1, f_2, \ldots, f_n)$ is connected by Theorem 2.4 being the image of $G'(f_2, f_3, \ldots, f_n)$ under the upper semicontinuous function ψ.   □

Theorem 2.6. *Suppose $\{X_1, X_2, \ldots, X_{n+1}\}$ is a finite collection of continua and $\{f_1, f_2, \ldots, f_n\}$ is a finite collection of upper semicontinuous functions such that $f_i : X_{i+1} \to C(X_i)$ for $1 \leq i \leq n$. Then, $G'(f_1, f_2, \ldots, f_n)$ is a continuum.*

Proof. If the finite collection of continua contains only one function f_1, then $G'(f_1) = (G(f_1))^{-1}$ is a continuum by Theorem 2.5.

Inductively, suppose $k \geq 2$ is an integer such that if $\{X_1, X_2, \ldots, X_k\}$ is a collection of k continua and $\{g_1, g_2, \ldots, g_{k-1}\}$ is a collection of $k - 1$ upper semicontinuous functions such that $g_i : X_{i+1} \rightarrow C(X_i)$ for $1 \leq i \leq k - 1$, then $G'(g_1, g_2, \ldots, g_{k-1})$ is a continuum. Let $\{X_1, X_2, \ldots, X_{k+1}\}$ be a collection of $k + 1$ continua and let $\{f_1, f_2, \ldots f_k\}$ be a collection of k upper semicontinuous functions such that $f_i : X_{i+1} \rightarrow C(X_i)$ for $1 \leq i \leq k$. By the inductive hypothesis, $G'(f_2, f_3, \ldots, f_k)$ is a continuum. It follows from Lemma 2.2 that $G'(f_1, f_2, \ldots, f_k)$ is a continuum.   □

We now prove the main theorem of this section.

Theorem 2.7. *Suppose X is a sequence of subintervals of $[0, 1]$ and f is a sequence of upper semicontinuous functions such that $f_i : X_{i+1} \rightarrow C(X_i)$. Then, $\varprojlim f$ is a continuum.*

Proof. Suppose $n \in \mathbb{N}$ and note that $G_n = G'(f_1, f_2, \ldots, f_n) \times \prod_{i>n+1} X_i$. By Theorem 2.6, $G'(f_1, f_2, \ldots, f_n)$ is a continuum, so G_n is a continuum being the product of two continua. That $\varprojlim f$ is a continuum now follows from Theorem 2.1.
                                                                                □

A stronger statement than that of Theorem 2.7 is true. We leave its proof to the interested reader. A proof in a very general setting may be found in [6, Theorem 125, p. 89].

Theorem 2.8. *Suppose X is a sequence of subintervals of $[0, 1]$ and f is a sequence of upper semicontinuous functions such that $f_i : X_{i+1} \rightarrow 2^{X_i}$. Suppose further that for each positive integer i such that f_i does not have connected values, $f_i(X_{i+1})$ is connected and $f_i^{-1}(x)$ is an interval for each $x \in f_i(X_{i+1})$. Then, $\varprojlim f$ is a continuum.*

By way of contrast to Theorem 2.7, in Example 2.7 below we see that an inverse limit can be connected even if most of the values of the bonding function are totally disconnected. This is only one of many such examples to be found in this book.

2.4 Examples

We consider some examples having connected inverse limits in which the connectedness is a consequence of Theorem 2.7. However, in these examples, we show more than that continua are produced in the inverse limit. We are able to say something about the nature of the inverse limit; in fact, in some cases, we identify what the inverse limit is and provide a model for it. Such is the case in our next example.

Example 2.2 (An arc). Let $f : [0, 1] \rightarrow C([0, 1])$ be given by $f(t) = 0$ for $0 \leq t < 1$ and $f(1) = [0, 1]$. Then $\varprojlim f$ is an arc (see the graph on the left side of Fig. 2.2 for $G(f)$ and Fig. 2.3 for a model of the inverse limit).

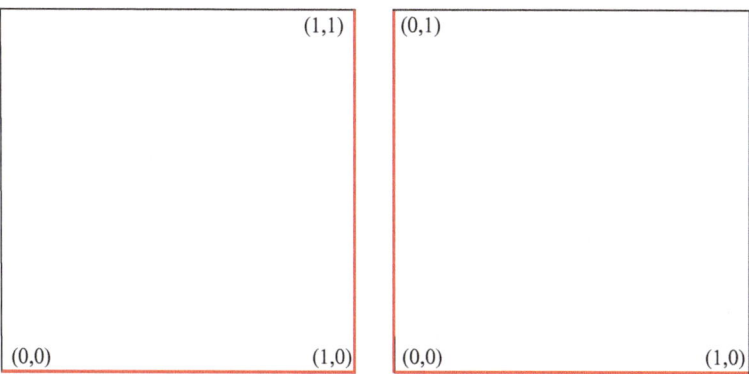

Fig. 2.2 Graphs of the bonding functions in Examples 2.2 (*left*) and 2.3 (*right*)

Fig. 2.3 A model of the arc that is the inverse limit in Example 2.2

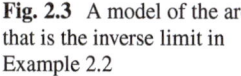
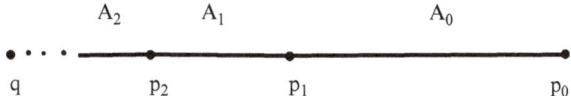

Proof. Let $M = \varprojlim f$. The connectedness of M follows from Theorem 2.7, but we wish to conclude that the inverse limit is an arc. To that end, let $q = (0, 0, 0, \dots)$, $p_0 = (1, 1, 1, \dots)$, and $A_0 = \{x \in M \mid x_i = 1 \text{ for } i > 1\}$. For each positive integer n, let p_n be the point whose first n coordinates are 0 and all remaining coordinates are 1 and let $A_n = \{x \in M \mid x_i = 0 \text{ for } i \leq n \text{ and } x_i = 1 \text{ for } i > n + 1\}$. Then, for each $n \geq 0$, A_n is an arc and $A_n \cap A_{n+1} = \{p_{n+1}\}$. It is not difficult to see that $M = (\bigcup_{n \geq 0} A_n) \cup \{(0, 0, 0, \dots)\}$, and if $x_l \subset A_l$ for each positive integer l, then $x_1, x_2, x_3, \dots$ converges to q. It follows that M is an arc. $\quad\square$

If we move the vertical line to the left side of $[0, 1]^2$ in Example 2.2, we get an entirely different inverse limit.

Example 2.3 (An infinite-dimensional continuum). Let $f : [0, 1] \to C([0, 1])$ be given by $f(0) = [0, 1]$ and $f(t) = 0$ for $0 < t \leq 1$. Then $\varprojlim f$ is an infinite-dimensional continuum (see the graph on the right side of Fig. 2.2 for $G(f)$).

Proof. The inverse limit is a continuum by Theorem 2.7, and it contains a continuum homeomorphic to $\mathcal{Q}$, namely, $([0, 1] \times \{0\})^\infty$. $\quad\square$

Placing the vertical line at $1/2$ yields yet a third continuum as we see in Example 2.4.

Example 2.4 (A "comb"). Let $f : [0, 1] \to C([0, 1])$ be given by $f(t) = 0$ for $t \neq 1/2$ and $f(1/2) = [0, 1]$. Then $\varprojlim f$ is the union of a sequence of arcs $A_1, A_2, A_3, \dots$ and the point $(0, 0, 0, \dots)$ such that, for each positive integer i, $A_{i+1} \cap A_i$ is a single point that is an endpoint of A_i and an interior point of A_{i+1} (see Fig. 2.4 for the graph of the bonding function and a model of the inverse limit).

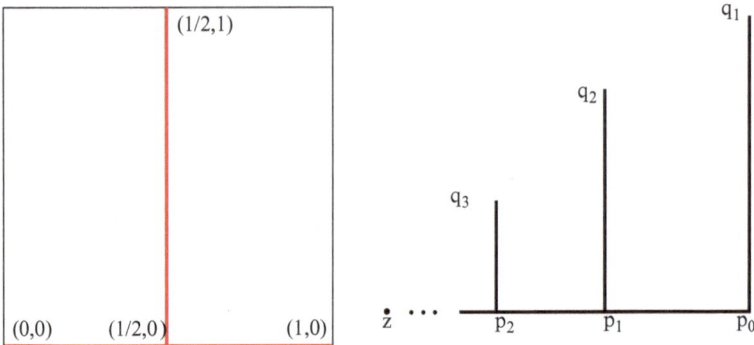

Fig. 2.4 The graph of the bonding function and a model of the inverse limit in Example 2.4

Proof. Let $M = \varprojlim f$ and, for each $i \in \mathbb{N}$, let $A_i = \{x \in M \mid x_i \in [0,1], x_j = 1/2$ for $j > i$, and if $i > 1$, then $x_j = 0$ for $j < i\}$. For each $i \in \mathbb{N}$, let p_i be the point of A_i such that $\pi_i(p_i) = 0$ and q_i be the point of A_i such that $\pi_i(q_i) = 1$. Then A_i is an arc with endpoints p_i and q_i. Letting $z = (0,0,0,\ldots)$, we have $\varprojlim f = \{z\} \cup (\bigcup_{n \geq 1} A_n)$. If $i \in \mathbb{N}$, then $A_i \cap A_{i+1} = \{p_i\}$. Let p_0 denote the point of M having every coordinate $1/2$. Note that p_i is an interior point of A_{i+1} for each nonnegative integer i. □

2.5 Topological Conjugacy

Placing the vertical line in Example 2.4 above any number other than 0 and 1 yields a continuum homeomorphic to the continuum of Example 2.4. To show this, we introduce the notion of topological conjugacy. Functions $f : [0,1] \to 2^{[0,1]}$ and $g : [0,1] \to 2^{[0,1]}$ are said to be *topologically conjugate* provided there is a surjective homeomorphism $h : [0,1] \twoheadrightarrow [0,1]$ such that $fh = hg$.

The following theorem may be found in [5, Theorem 5.3, p. 126] in a more general setting.

Theorem 2.9. *If $f : [0,1] \to 2^{[0,1]}$ and $g : [0,1] \to 2^{[0,1]}$ are topologically conjugate upper semicontinuous functions, then $\varprojlim f$ and $\varprojlim g$ are homeomorphic.*

Proof. Let $M = \varprojlim f$ and $N = \varprojlim g$. If $h : [0,1] \twoheadrightarrow [0,1]$ is a surjective homeomorphism such that $fh = hg$, then $H : N \twoheadrightarrow M$ given by $H(x) = (h(x_1), h(x_2), h(x_3), \ldots)$ is a homeomorphism from N onto M. To see this, note that the function H is continuous because it is coordinatewise continuous and it is 1–1 because h is 1–1. That $H(x) \in M$ for each $x \in N$ is a consequence of $fh = hg$ for $h(x_i) \in h(g(x_{i+1})) = f(h(x_{i+1}))$.

Fig. 2.5 The graph of the
bonding function in
Example 2.6

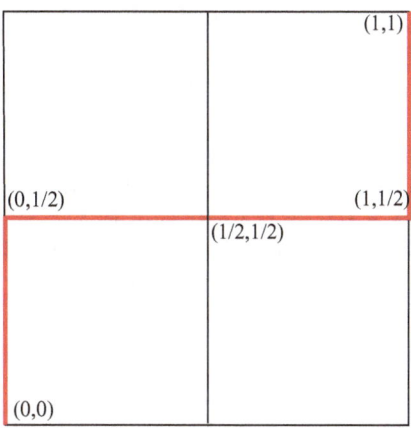

If $y \in M$, then $x = (h^{-1}(y_1), h^{-1}(y_2), h^{-1}(y_3), \dots)$ is a point of Q such that $H(x) = y$. That $x \in N$ may be seen as follows. Because $f = hgh^{-1}$ and $y \in M$, for each positive integer i, $y_i \in hgh^{-1}(y_{i+1})$. Therefore, $h^{-1}(y_i) \in g(h^{-1}(y_{i+1}))$, i.e., $x_i \in g(x_{i+1})$.   □

Example 2.5. If $0 < c < 1$ and $f : [0, 1] \to C([0, 1])$ is the function given by $f(t) = 0$ for $t \neq c$ and $f(c) = [0, 1]$, then $\underleftarrow{\lim} f$ is homeomorphic to the inverse limit from Example 2.4.

Proof. Denote by g the function from Example 2.4. Using Theorem 2.9, it can be seen that $\underleftarrow{\lim} f$ is homeomorphic to $\underleftarrow{\lim} g$. Indeed, the piecewise linear homeomorphism h whose graph is the union of two straight line intervals, one from $(0, 0)$ to $(1/2, c)$ and the other from $(1/2, c)$ to $(1, 1)$, is a conjugacy because $fh = hg$.   □

For $c \in [0, 1]$, let $g_c : [0, 1] \to C([0, 1])$ denote the function given by $g_c(t) = 1$ for $t \neq c$ and $g_c(c) = [0, 1]$. Because the function f from Example 2.5 is conjugate to the function g_{1-c} under the homeomorphism $1 - Id$, $\underleftarrow{\lim} f$ is homeomorphic to $\underleftarrow{\lim} g_{1-c}$. For $c = 0$, the function g_c is conjugate under $1 - Id$ to the bonding function in Example 2.2 so $\underleftarrow{\lim} g_c$ is an arc; for $c = 1$, the function g_c is conjugate under $1 - Id$ to the bonding function in Example 2.3, so $\underleftarrow{\lim} g_c$ is infinite dimensional.

We end this section with an additional application of Theorem 2.9. Our proof also uses Theorem 1.9.

Example 2.6 (An arc). Let $f : [0, 1] \to C([0, 1])$ be given by $f(0) = [0, 1/2]$, $f(t) = 1/2$ for $0 < t < 1$, and $f(1) = [1/2, 1]$ (see Fig. 2.5 for the graph of f). Then, $\underleftarrow{\lim} f$ is an arc.

Proof. Let $f_1 = f|[0, 1/2]$ and $f_2 = f|[1/2, 1]$. Because $G(f_i) \subseteq G(f)$, $\underleftarrow{\lim} f_i \subseteq \underleftarrow{\lim} f$ for $i = 1, 2$. Suppose $x \in \underleftarrow{\lim} f$ and $x \neq (1/2, 1/2, 1/2, \dots)$. Let k be the least integer j such that $x_j \neq 1/2$. If $x_k < 1/2$, then $x_j = 0$ for

$j \geq k$, and, in case $k > 1$, $x_j = 1/2$ for $j < k$. If $x_k > 1/2$, then $x_k = 1$ for $j \geq k$, and, in case $k > 1$, $x_j = 1/2$ for $j < k$. Thus, $x \in \lim \overleftarrow{f_1}$ when $x_k < 1/2$ and $x \in \lim \overleftarrow{f_2}$ when $x_k > 1/2$. Therefore, $\lim \overleftarrow{f} = \lim \overleftarrow{f_1} \cup \lim \overleftarrow{f_2}$. Let $g : [0, 1] \to C([0, 1])$ be the function given by $g(t) = 0$ for $0 \leq t < 1$ and $g(1) = [0, 1]$ (i.e., g is the function from Example 2.2 having an arc as its inverse limit). Then f_1 and g are conjugate under the homeomorphism h_1 given by $h_1(t) = 1/2 - t/2$ for $0 \leq t \leq 1$, while f_2 and g are conjugate under the homeomorphism h_2 given by $h_2(t) = 1/2 + t/2$ for $0 \leq t \leq 1$. Thus, $\lim \overleftarrow{f}$ is the union of two arcs intersecting only at $\{(1/2, 1/2, 1/2, \ldots)\}$ and is therefore an arc.

$\square$

2.6 Connectedness from Bonding Functions Without Connected Values

An inverse limit with a single surjective bonding function can be connected even if the function does not have all of its values connected. We shall see many such examples in this book. Example 2.7 below is an interesting one. The inverse limit is the cone over the Cantor set, often called the *Cantor fan*. There are a Cantor set C lying in the inverse limit and a point v of the inverse limit such that the inverse limit is the union of a collection of arcs each having v as one of its endpoints with its other endpoint in C and such that the only point common to any two of the arcs is v.

Example 2.7 (The Cantor fan). Let $f : [0, 1] \to 2^{[0,1]}$ be given by $f(t) = \{t, 1-t\}$ for $0 \leq t \leq 1$ (i.e., $G(f)$ is the union of Id and $1 - Id$). Then $\lim \overleftarrow{f}$ is the Cantor fan with vertex $v = (1/2, 1/2, 1/2, \ldots)$. (See Fig. 2.6.)

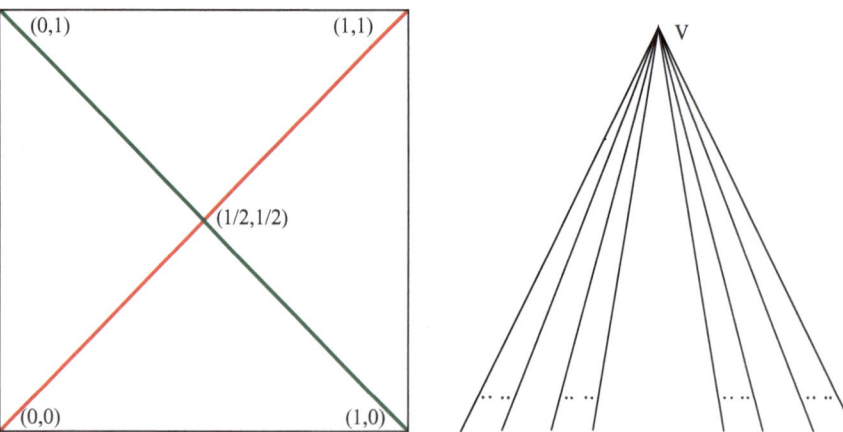

Fig. 2.6 The graph of the bonding function and a model of the inverse limit in Example 2.7

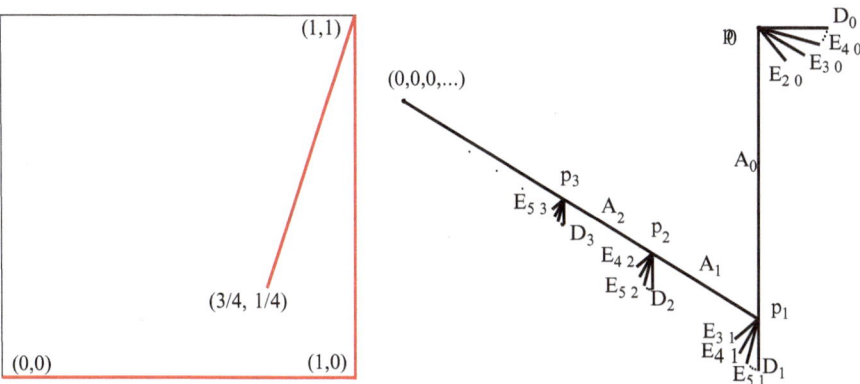

Fig. 2.7 The graph of the bonding function and a model of the inverse limit in Example 2.8

Proof. There are four homeomorphisms whose union is f. They are $g_1 : [0, 1/2] \to$ $[0, 1/2]$ given by $g_1(t) = t$, $g_2 : [0, 1/2] \to [1/2, 1]$ given by $g_2(t) = 1 - t$, $g_3 :$ $[1/2, 1] \to [0, 1/2]$ given by $g_3(t) = 1 - t$, and $g_4 : [1/2, 1] \to [1/2, 1]$ given by $g_4(t) = t$. A point $\boldsymbol{x}$ is in $\varprojlim \boldsymbol{f}$ if and only if there is a sequence $h_1, h_2, h_3, \ldots$ such that $h_i \in \{g_1, g_2, g_3, g_4\}$ for each i and $x_i = h_i(x_{i+1})$ for each positive integer i. Each such inverse limit is an arc having $\boldsymbol{v} = (1/2, 1/2, 1/2, \ldots)$ as one endpoint and the other endpoint in the Cantor set $\{0, 1\}^\infty$. Moreover, if $\boldsymbol{p} \in \{0, 1\}^\infty$, there is a sequence $\boldsymbol{h}$ such $h_i \in \{g_1, g_2, g_3, g_4\}$ for each i and $\boldsymbol{p} \in \varprojlim \boldsymbol{h}$. □

The connectedness of Example 2.7 is demonstrated here in an ad hoc manner, although it is a consequence of a theorem that we prove later (see Theorem 2.11). We now include some additional examples of inverse limits that are continua even though in each case, the bonding function does not have all of its values connected. The connectedness of each of the next two examples is also demonstrated in an ad hoc manner. Unlike the previous example, however, we do not have a subsequent theorem from which the connectedness follows (see Problem 6.5). We make additional use of these examples later.

Example 2.8. Let $f : [0, 1] \to 2^{[0,1]}$ be the function given by $f(t) = 0$ for $0 \le t < 3/4$, $f(t) = \{0, 3t - 2\}$ for $3/4 \le t < 1$, and $f(1) = [0, 1]$. Then $\varprojlim \boldsymbol{f}$ is a continuum. (See Fig. 2.7 for the graph of f and a model of its inverse limit.)

Proof. Let $M = \varprojlim \boldsymbol{f}$. Let g be the bonding function from Example 2.2, i.e., $g : [0, 1] \to C([0, 1])$ is given by $g(t) = 0$ for $0 \le t < 1$ and $g(1) = [0, 1]$, and let $A = \varprojlim \boldsymbol{g}$. Then A is an arc and $A \subseteq M$ because $G(g) \subseteq G(f)$. Let $\boldsymbol{p_0}$ be the point $(1, 1, 1, \ldots)$ and, for each positive integer j, let $\boldsymbol{p_j}$ be the point of M whose first j coordinates are 0 and all other coordinates are 1. Each point of the sequence $\boldsymbol{p_0}, \boldsymbol{p_1}, \boldsymbol{p_2}, \ldots$ is a point of A. For $j \ge 0$, let $D_j = \{\boldsymbol{x} \in M \mid 1/4 \le x_{j+1} \le 1, x_{k+1} = (x_k + 2)/3$ for $k > j$, and, if $j > 0, x_k = 0$ for $1 \le k \le j\}$. For each

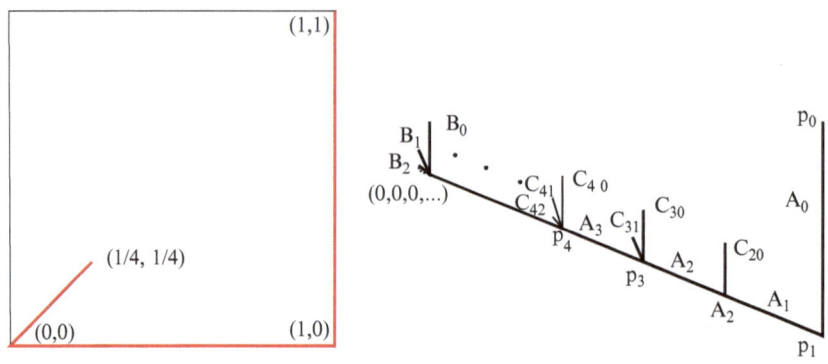

Fig. 2.8 The graph of the bonding function and a model of the inverse limit in Example 2.9

integer $j \geq 0$ and each integer i such that $i \geq j + 2$, let $E_{ij} = \{x \in M \mid 1/4 \leq x_{j+1} \leq 1, x_{k+1} = (x_k + 2)/3 \text{ for } j + 1 \leq k < i, x_k = 1 \text{ for } k > i \text{ and, if } j > 0, x_k = 0 \text{ for } 1 \leq k \leq j\}$. For each $n \geq 0$, $F_n = D_n \cup (\bigcup_{k>n+1} E_{kn})$ is a fan with vertex p_n. Note that $M = A \cup (\bigcup_{n \geq 0} F_n)$, so M is connected. □

Example 2.9. Let $f : [0, 1] \to 2^{[0,1]}$ be the function given by $f(t) = \{0, t\}$ for $0 \leq t \leq 1/4$, $f(t) = 0$ for $1/4 < t < 1$, and $f(1) = [0, 1]$. Then, $\varprojlim f$ is a continuum. (See Fig. 2.8 for the graph of f and a model of its inverse limit.)

Proof. Let $M = \varprojlim f$. As in the previous example, let g be the bonding function from Example 2.2, i.e., $g : [0, 1] \to C([0, 1])$ is given by $g(t) = 0$ for $0 \leq t < 1$ and $g(1) = [0, 1]$, and let $A = \varprojlim g$. Then A is an arc and $A \subseteq M$ because $G(g) \subseteq G(f)$. Let i and j be integers with $i \geq 2$ and $0 \leq j < i - 1$. Let $C_{ij} = \{x \in M \mid x_i \in [0, 1/4], x_k = x_i \text{ for } j < k \leq i, x_k = 1 \text{ for } k > i,$ and if $j > 0, x_k = 0 \text{ for } 1 \leq k \leq j\}$. Let $B_0 = \{x \in M \mid x_k \in [0, 1/4] \text{ and } x_{k+1} = x_k \text{ for each positive integer } k\}$ and, for each positive integer i, let $B_i = \{x \in M \mid x_{i+1} \in [0, 1/4] \text{ and } x_k = x_{i+1} \text{ for } k \geq i + 1 \text{ and } x_k = 0 \text{ for } k \leq i\}$. Note that $F = \bigcup_{i \geq 0} B_i$ is a fan with vertex $(0, 0, 0, \ldots)$, its point of intersection with A. Let $p_0 = (1, 1, 1, \ldots)$ and, for each positive integer i, let p_i be the point such that the first i coordinates of p_i are 0 and the remaining coordinates are 1. If i and j are integers with $i \geq 2$ and $0 \leq j < i - 1$, then C_{ij} intersects A at the point p_i. To see that M is connected, one only need observe that if $x \in M - (A \cup F)$, then x is in C_{ij} for some i, j. □

We close this section with a simple theorem that is easy to prove. Except for the case that some of the terms of the sequence of mappings in the hypothesis of Theorem 2.10 are homeomorphisms, the bonding functions do not have all of their values connected.

Theorem 2.10. *If g is a sequence of surjective mappings of $[0, 1]$ onto $[0, 1]$ and $f_i = g_i^{-1}$ for each $i \in \mathbb{N}$, then $\varprojlim f$ is an arc.*

Proof. Because g_i is surjective, $f_i : [0,1] \to 2^{[0,1]}$ is an upper semicontinuous function for each $i \in \mathbb{N}$. Then, $h : [0,1] \to \varprojlim f$ given by $h(t) = (t, g_1(t), g_2(g_1(t)), g_3(g_2(g_1(t))), \ldots)$ is a homeomorphism of $[0,1]$ onto $\varprojlim f$. $\square$

2.7 Union Theorems

One method of obtaining connected inverse limits with set-valued functions is to use upper semicontinuous bonding functions having graphs that are set-theoretic unions of the graphs of upper semicontinuous functions with connected values as shown below in Theorem 2.11. Because mappings (continuous functions) on $[0,1]$ have connected values, set-valued functions that are unions of mappings often (but not always) produce connected inverse limits. That some restrictions along the lines of those in Theorem 2.11 must be imposed can be seen from Example 1.2 where the bonding function is the union of two constant maps and the inverse limit is a Cantor set.

Our first theorem in this section is due to Nall [Theorem 3.1, 10], although we have cast it in slightly different language from his original statement. Theorem 2.11 generalizes a theorem on unions of upper semicontinuous continuum-valued functions published earlier [3, Theorem 2.12, p. 363] (at least in the metric setting), and its hypothesis is perhaps somewhat easier to verify. After proving this theorem, we provide examples that can be shown to be connected using it. Although Nall proves this theorem for compact metric spaces, we state and prove it on $[0,1]$. Recall that if $\{X_1, X_2, \ldots, X_{n+1}\}$ is a finite collection of closed subsets of $[0,1]$ and $\{f_1, f_2, \ldots, f_n\}$ is a finite collection of functions such that $f_i : X_{i+1} \to 2^{X_i}$ for $1 \le i \le n$ and $G'(f_1, f_2, \ldots, f_n) = \{x \in \prod_{i=1}^{n+1} X_i \mid x_i \in f_i(x_{i+1}) \text{ for } 1 \le i \le n\}$.

Theorem 2.11 (Nall). *Suppose $\mathcal{F}$ is a collection of upper semi-continuous functions such that if $g \in \mathcal{F}$, then $g : [0,1] \to C([0,1])$, and f is the function whose graph is the set-theoretic union of all of the graphs of the functions in $\mathcal{F}$. If f is surjective and $G(f)$ is a continuum, then $\varprojlim f$ is a continuum.*

Proof. Because $G(f)$ is a continuum, f is upper semicontinuous. By Theorem 2.1, showing that $G_n = \{x \in \mathcal{Q} \mid x_i \in f(x_{i+1}) \text{ for } 1 \le i \le n\}$ is connected for each $n \in \mathbb{N}$ is sufficient to prove the theorem. To that end, we proceed by induction.

Note that $G_1 = G(f^{-1}) \times \mathcal{Q}$ is connected, being the product of two connected sets, so G_1 is connected.

Suppose k is a positive integer such that G_k is connected. We adopt the following notation. If j is a positive integer, let $G'_j = G'(f_1, f_2, \ldots, f_j)$ where $f_i = f$ for $1 \le i \le j$. Then, $G'_k = p_{\{1,2,\ldots,k+1\}}(G_k)$ is connected. Suppose G'_{k+1} is the union of two closed sets H and K. Then, $G(f^{-1}) = p(G'_{k+1}) = p(H) \cup p(K)$ where $p : [0,1]^{k+2} \to [0,1]^2$ is the mapping given by $p(x) = (x_1, x_2)$. Because $G(f)$ is connected, $G(f^{-1})$ is connected, so there is a point $(c,d) \in p(H) \cap p(K)$.

Fig. 2.9 The graph of
bonding the function in
Example 2.10

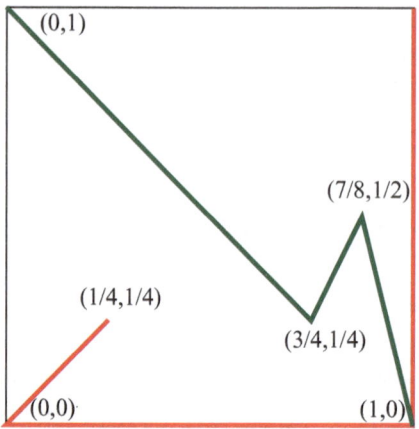

There are points $x \in H$ and $y \in K$ such that $p(x) = (c, d) = p(y)$. There is
a function $g \in \mathcal{F}$ such that $(d, c) \in g$. By Lemma 2.2, $G'(g_1, g_2, \ldots, g_{k+1})$ is a
connected subset of G'_{k+1} where $g_1 = g$ and $g_i = f$ for $2 \leq i \leq k + 1$ and
contains both x and y. Thus, $H \cap K \neq \emptyset$, and it follows that G'_{k+1} is connected.
Because $G_{k+1} = G'_{k+1} \times Q$, G_{k+1} is connected.   □

A major reason for at least some of the initial interest in inverse limits with upper
semicontinuous functions of the type satisfying the hypothesis of Theorem 2.11 is a
potential application to economics. Models in backward economics can involve two
mappings, and it is important to understand the potential outcomes of the models
no matter which of the mappings is used at each stage. Thus, we are led to consider
inverse limits with set-valued functions having graphs that are the union of two
mappings. However, this topic is of interest in its own right because of theorems
like Theorem 2.11. Indeed, most of the research on set-valued functions that are
unions of mappings has been concentrated on determining when the inverse limit
is a continuum, and this is the case for two maps with a coincidence point and
a surjective union. It would be of interest to conduct a study of inverse limits of
upper semicontinuous functions that are the union of two maps of $[0, 1]$ that do not
have a coincidence point (see Problem 6.6) even though such inverse limits are not
connected.

Our next example demonstrates that we cannot weaken the hypothesis in
Theorem 2.11 that the elements of $\mathcal{F}$ have connected values to require simply that
the elements of $\mathcal{F}$ have connected inverse limits.

Example 2.10. Let $f_1 : [0, 1] \to 2^{[0,1]}$ be given by $f_1(t) = \{0, t\}$ for $0 \leq t \leq 1/4$,
$f_1(t) = 0$ for $1/4 < t < 1$, and $f_1(1) = [0, 1]$. Let $g : [0, 1] \twoheadrightarrow [0, 1]$ be the
mapping given by $g(t) = 1 - t$ for $0 \leq t \leq 3/4$, $g(t) = 2t - 5/4$ for $3/4 < t \leq 7/8$,
and $g(t) = 4 - 4t$ for $7/8 < t \leq 1$. Let f be the upper semicontinuous function
whose graph is $G(f_1) \cup g$. Then $\varprojlim f_1$ and $\varprojlim g$ are connected, but $\varprojlim f$ is not
connected. (See Fig. 2.9 for a graph of f.)

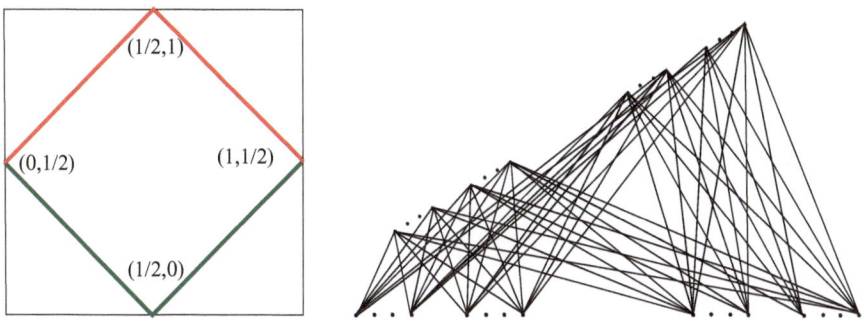

Fig. 2.10 The graph of the bonding function and a model of the inverse limit in Example 2.11

Proof. Let $M = \varprojlim f$. Because f_1 is the bonding function from Example 2.9, its inverse limit is connected. Because g is a mapping, its inverse limit is connected. Let $N = \{x \in M \mid x_1 = x_2 = 1/4 \text{ and } x_3 = 3/4\}$ and note that N is closed. However, because $N = M \cap ((1/8, 3/8) \times (1/8, 3/8) \times (5/8, 7/8) \times \mathcal{Q})$, N is also open in M. Thus, M is not connected.   □

Example 2.11. Let $g_1 : [0, 1] \rightarrow [0, 1]$ be the mapping given by $g_1(t) = t + 1/2$ for $0 \le t \le 1/2$ and $g_1(t) = 3/2 - t$ for $1/2 \le t \le 1$. Let $g_2 : [0, 1] \rightarrow [0, 1]$ be the mapping given by $g_2(t) = 1/2 - t$ for $0 \le t \le 1/2$ and $g_2(t) = t - 1/2$ for $1/2 \le t \le 1$. Let $\mathcal{F} = \{g_1, g_2\}$ and $f : [0, 1] \rightarrow 2^{[0,1]}$ be the upper semicontinuous function whose graph is the set-theoretic union of g_1 and g_2. Then, $\varprojlim f$ is a nonplanar continuum. (See Fig. 2.10 for the graph of f and a model of the inverse limit.)

Proof. Let $M = \varprojlim f$. Because g_1 and g_2 are mappings, $G(f) = g_1 \cup g_2$ is connected, and f is surjective; the proof that M is a continuum is a simple application of Theorem 2.11.

To obtain a model for the inverse limit, we view it in a slightly different way. There are two intervals $J_1 = [0, 1/2]$ and $J_2 = [1/2, 1]$ and four mappings $f_1 : J_1 \twoheadrightarrow J_1$, $f_2 : J_2 \twoheadrightarrow J_1$, $f_3 : J_2 \twoheadrightarrow J_2$, and $f_4 : J_1 \twoheadrightarrow J_2$ such that $G(f) = f_1 \cup f_2 \cup f_3 \cup f_4$. The continuum M contains two Cantor sets: C_1 containing all the points p of M with all odd coordinates $1/2$ and all even coordinates in $\{0, 1\}$ and C_2 containing all the points p of M with all even coordinates $1/2$ and all odd coordinates in $\{0, 1\}$. The continuum M consists of all arcs of the form $\varprojlim g$ where, for each $i \in \mathbb{N}$, $g_i \in \{f_1, f_2, f_3, f_4\}$ and the domain of g_i is the range of g_{i+1}. Each such arc joins a point of C_1 with a point of C_2, and furthermore, if p is point of C_1 and q is a point of C_2, there is such an arc having endpoints p and q. Moreover, each two such arcs that intersect do so at only one point belonging to $C_1 \cup C_2$. The reader should note that M contains numerous simple closed curves. Because each point of C_1 is a vertex of a Cantor fan over C_2 and each two such Cantor fans contain mutually exclusive fans, M contains uncountably many mutually exclusive triods

Fig. 2.11 The graph of the
bonding function in
Example 2.12

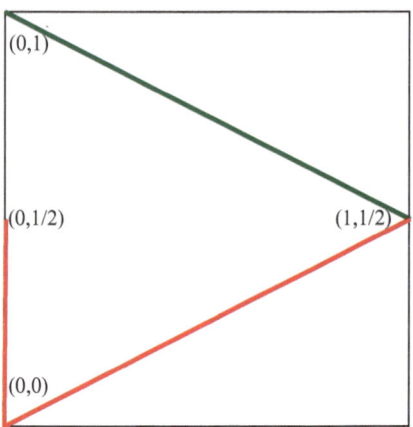

and so is nonplanar [8, Theorem 84, p. 222] (a *triod* is a continuum that contains a
subcontinuum having a complement with at least three components).

The continuum can also be seen to be nonplanar because it contains a Kuratowski
complete bipartite graph $K_{3,3}$ that consists of six vertices, three from C_1 and three
from C_2, and edges joining each vertex of the three in C_1 with each vertex of the
three in C_2.

To depict our model, choose two skew lines in three-dimensional Euclidean space
and embed C_1 in one of these lines and C_2 in the other. By joining each point of C_1
with each point of C_2 by a straight line interval, we obtain a model of M.   □

The inverse limit M in Example 2.11 is the well-known *Hurewicz continuum*
having the property that if C is a continuum, then there exist a subcontinuum H of
M and a monotone mapping of H onto C [2].

We end this section with one more example of an inverse limit that we show is
a continuum using Theorem 2.11. The function in this example does not satisfy the
hypothesis of Theorem 2.12 of [3].

Example 2.12. Let T denote the full tent map, $T(t) = 2t$ for $0 \le t \le 1/2$ and
$T(t) = 2 - 2t$ for $1/2 \le t \le 1$. Let $f : [0, 1] \to 2^{[0,1]}$ be the upper semicontinuous
function whose graph $G(f) = T^{-1} \cup (\{0\} \times [0, 1/2])$. Then $\varprojlim f$ is a continuum.
(See Fig. 2.11 for the graph of f and Fig. 2.12 for an indication of a model for the
inverse limit.)

Proof. Let $g_1 : [0, 1] \to [0, 1]$ be given by $g_1(t) = 1 - t/2$ for $0 \le t \le 1$ and
$g_2 : [0, 1] \to C([0, 1])$ be given by $g_2(0) = [0, 1/2]$ and $g_2(t) = t/2$ for $0 < t \le 1$.
Because $G(f) = G(g_1) \cup G(g_2)$ is connected, the proof that $M = \varprojlim f$ is a
continuum is a simple application of Theorem 2.11.

We now construct a model for this inverse limit. Let $\varphi : [0, 1] \to [0, 1]$ be the
homeomorphism given by $\varphi(t) = t/2$ and denote by $\hat{\varphi}$ the mapping of M given by
$\varphi(\boldsymbol{x}) = (\varphi(x_1), x_1, x_2, \dots)$. Because φ is a homeomorphism and $\varphi \subseteq G(f)$, $\hat{\varphi}$ is

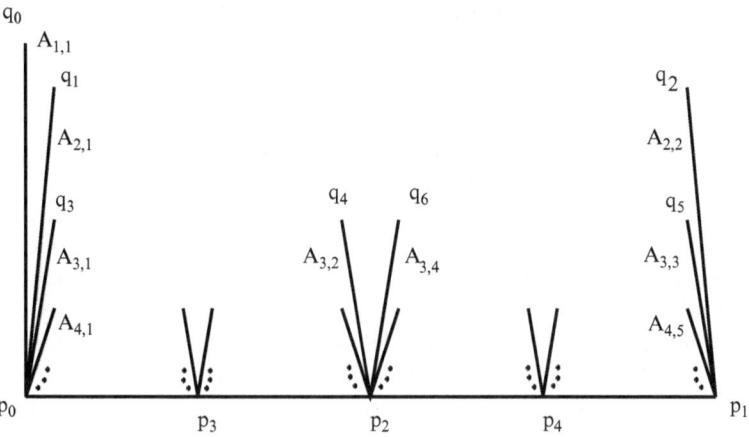

Fig. 2.12 A model of the inverse limit in Example 2.12

a homeomorphism of M into M. Let $\psi : [0, 1] \to [0, 1]$ be the homeomorphism given by $\psi(t) = 1 - t/2$ and denote by $\hat{\psi}$ the homeomorphism of M into M given by $\psi(x) = (\psi(x_1), x_1, x_2, \dots)$.

Let $A = \varprojlim T^{-1}$. By Theorem 1.9, A is a subset of M because $G(T^{-1}) \subseteq G(f)$. By Theorem 2.10, A is an arc. The endpoints of A are $p_0 = (0, 0, 0, \dots)$ and $p_1 = (1, 0, 0, \dots)$.

Suppose $x \in M$ and $x \notin A$. There is a positive integer n such that $x_n \in (0, 1/2]$ and $x_j = 0$ for $j > n$. Let $M_n = \{x \in M \mid x_n \in [0, 1/2] \text{ and } x_j = 0 \text{ for } j > n\}$. It follows that $M = A \cup (\bigcup_{i > 0} M_i)$.

As we proceed, we identify arcs and points that are shown in our model of M depicted in Fig. 2.12. With that in mind, note that M_1 is an arc with endpoints p_0 and $q_0 = (1/2, 0, 0, 0, \dots)$. Letting $A_{1,1} = M_1$, we see that $M_2 = \hat{\varphi}(A_{1,1}) \cup \hat{\psi}(A_{1,1})$. Let $A_{2,1} = \hat{\varphi}(A_{1,1})$ and $A_{2,2} = \hat{\psi}(A_{1,1})$. So, $A_{2,1}$ is an arc with endpoints $p_0 = (0, 0, 0, \dots)$ and $q_1 = (1/4, 1/2, 0, 0, \dots)$, while $A_{2,2}$ is an arc with endpoints $p_1 = (1, 0, 0, 0, \dots)$ and $q_2 = (3/4, 1/2, 0, 0, \dots)$. Further, M_3 is the union of four arcs $A_{3,1} = \hat{\varphi}(A_{2,1})$, $A_{3,2} = \hat{\varphi}(A_{2,2})$, $A_{3,3} = \hat{\psi}(A_{2,1})$, and $A_{3,4} = \hat{\psi}(A_{2,2})$. Note that the endpoints of $A_{3,1}$ are p_0 and $q_3 = (1/8, 1/4, 1/2, 0, 0, \dots)$; the endpoints of $A_{3,2}$ are $p_2 = (1/2, 1, 0, 0, 0, \dots)$ and $q_4 = (3/8, 3/4, 1/2, 0, 0, \dots)$; $A_{3,3}$ has endpoints p_1 and $q_5 = (7/8, 1/4, 1/2, 0, 0, \dots)$; and $A_{3,4}$ has endpoints p_2 and $q_6 = (5/8, 3/4, 1/2, 0, 0, \dots)$. Continuing inductively, we observe that, for each positive integer n, M_{n+1} is the union of 2^n arcs that are, respectively, the images under $\hat{\varphi}$ of the arcs that comprise the components of M_n together with the arcs that are the images under $\hat{\psi}$ of the arcs comprising the components of M_n. □

We revisit Example 2.12 in Chap. 4 (Example 4.3).

2.8   Examples from Eight Similar Functions

In this section we consider inverse limits produced by eight similar graphs of upper semicontinuous functions that one obtains by the following process: choose one of the corners of $[0, 1] \times [0, 1]$ and take the union of the diagonal of the square emanating from that point and either the horizontal or the vertical side of the square that emanates from that point. Use that union as the graph of an upper semicontinuous set-valued function. Let $\mathcal{E}$ denote the collection of these eight inverse limits. Due to the fact that these eight graphs consist of the graphs of four topologically conjugate pairs of upper semicontinuous functions, we may examine four such graphs and through Theorem 2.9 know all the elements of $\mathcal{E}$. Interestingly enough, $\mathcal{E}$ contains four quite different continua even though the graphs that produce them are very similar. We begin with perhaps the simplest of these examples by choosing the graph that is the union of the diagonal and the horizontal side of the square lying on the bottom of the square.

Example 2.13 (A simple fan). Let $f : [0, 1] \rightarrow C([0, 1])$ be given by $f(t) = \{0, t\}$ for $0 \leq t \leq 1$. Then, $\varprojlim f$ is a fan with vertex $v = (0, 0, 0, \dots)$. (See Fig. 2.13 for the graph of f and a model of its inverse limit.)

Proof. Let $M = \varprojlim f$. That M is a continuum is a consequence of Theorem 2.8 (or of Theorem 2.11 by observing that $G(f)$ is a union of two mappings). Let $A_0 = \{x \in M \mid x_j = x_1 \text{ for each positive integer } j\}$. If $i \in \mathbb{N}$, let $A_i = \{x \in M \mid x_j = 0 \text{ for } j \leq i \text{ and } x_j = x_{i+1} \text{ for } j > i + 1\}$. Note that A_i is an arc containing $v = (0, 0, 0, \dots)$ for each nonnegative integer i. Moreover, $\varprojlim f = \bigcup_{k \geq 0} A_k$. $\square$

The function in Example 2.13 is conjugate to the function $g : [0, 1] \rightarrow 2^{[0,1]}$ given by $g(t) = \{t, 1\}$ for $0 \leq t \leq 1$ under the homeomorphism $h = 1 - Id$. Consequently, $\varprojlim g$ is homeomorphic to the inverse limit from Example 2.13.

Example 2.14. Let $f : [0, 1] \rightarrow C([0, 1])$ be given by $f(t) = t$ for $0 \leq t < 1$ and $f(1) = [0, 1]$. Then the inverse limit is a fan with vertex $v = (1, 1, 1, \dots)$. (See Fig. 2.14 for the graph of f and a model of its inverse limit.)

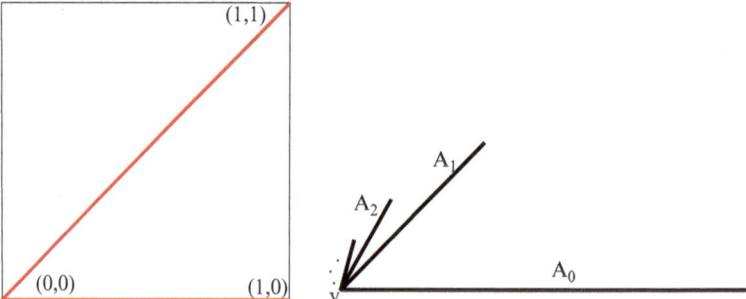

Fig. 2.13 The graph of the bonding function and a model of the inverse limit in Example 2.13

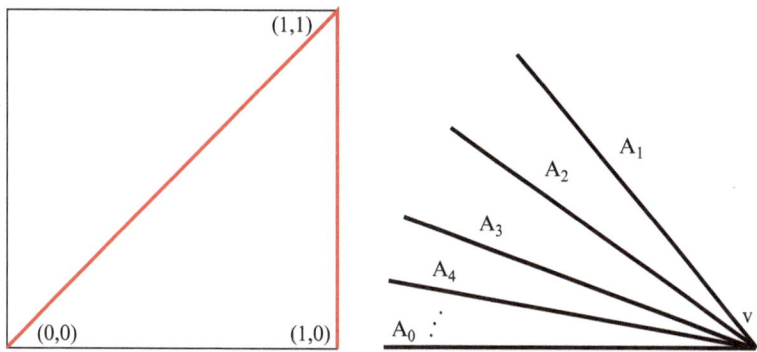

Fig. 2.14 The graph of the bonding function and a model of the inverse limit in Example 2.14

Proof. Let $M = \varprojlim f$ and let $A_0 = \{x \in M \mid x_j = x_1 \text{ for each } j\}$. That M is a continuum is a consequence of Theorem 2.7. Let i be a positive integer and let $A_i = \{x \in M \mid x_j = 1 \text{ for } j > i\}$. Note that A_k is an arc containing $v = (1, 1, 1, \dots)$ for each positive integer k. Moreover, $\varprojlim f = \bigcup_{k>0} A_k$. □

The function in Example 2.14 is conjugate to the function $g : [0, 1] \to C([0, 1])$ given by $g(0) = [0, 1]$ and $g(t) = t$ for $0 < t \le 1$ under the homeomorphism $h = 1 - Id$. Consequently, $\varprojlim g$ is homeomorphic to the inverse limit from Example 2.14.

It is known that if $0 < c < 1$ and $f_c : [0, 1] \twoheadrightarrow [0, 1]$ is the mapping whose graph is the union of two straight line intervals one from $(0, 0)$ to $(c, 1)$ and the other from $(c, 1)$ to $(1, 0)$, then $\varprojlim f_c$ is homeomorphic to the BJK horseshoe, $\varprojlim f_c$ for $c = 1/2$ (i.e., f_c is the full tent map). In Example 2.14, we examined the corresponding set-valued function for $c = 1$. Although we do not get a fan as an inverse limit of the upper semicontinuous set-valued function for $c = 0$, we next look at the surprisingly complicated inverse limit for this function.

Example 2.15. Let $f : [0, 1] \to C([0, 1])$ be given by $f(0) = [0, 1]$ and $f(t) = 1-t$ for $0 < t \le 1$. The complicated inverse limit $\varprojlim f$ is a nonplanar continuum that contains numerous $\sin(1/x)$-curves, two copies of the inverse limit from Example 2.14 attached along the limit arc, and many mutually exclusive n-ods for each positive integer n. (See Fig. 2.15 for the graph of f and Fig. 2.16 for a model of the inverse limit.)

Proof. Let $M = \varprojlim f$; M is a continuum by Theorem 2.7. This continuum is reasonably simple to describe, although it is rather complicated in its nature. Let $A = \{x \in M \mid x_{j+1} = 1 - x_j \text{ for each positive integer } j\}$. For each positive integer n, let $B_n = \{x \in M \mid x_{n+1} = 0\}$. Then, A is an arc, and each B_n is a product of an arc with a Cantor set. Note that $M = A \cup (\bigcup_{i>0} B_i)$. Denote by $\mathcal{B}_n$ the collection of arcs that are the components of B_n.

Fig. 2.15 The graph of the
bonding function in
Example 2.15

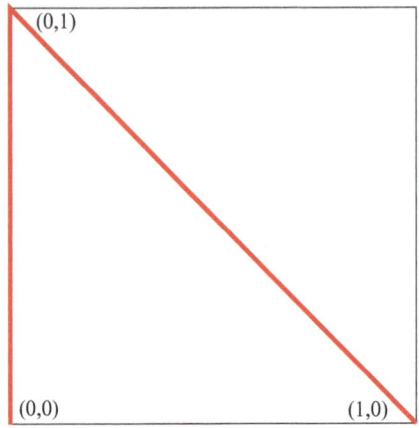

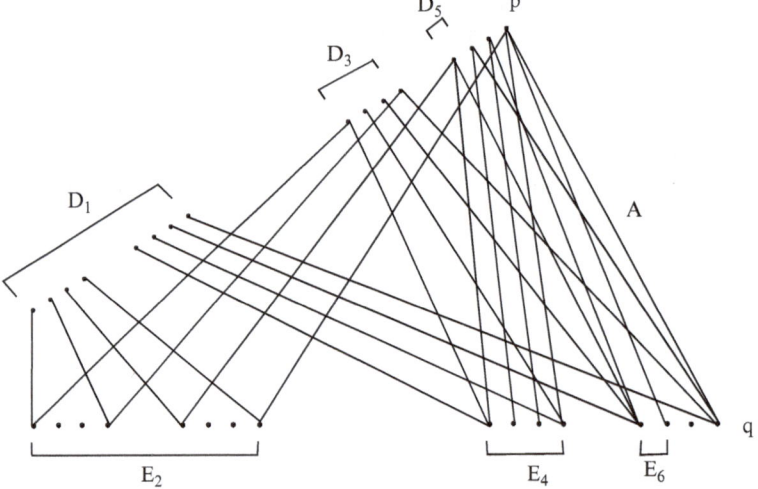

Fig. 2.16 A model of the inverse limit in Example 2.15

To obtain a model for M, we provide the following description of M that also
allows us to indicate the properties listed for it. There is a Cantor set C lying in
M that results from $\varprojlim g$ where $g(0) = \{0, 1\}$ and $g(1) = 0$. Let $C_0 = \{x \in C \mid x_1 = 0\}$, $C_1 = \{x \in C \mid x_1 = 1\}$. Then, $C = C_0 \cup C_1$. We partition C_0 and
C_1 in the following way. For each positive integer n, let p_n be the point of $\{0, 1\}^n$
such that $\pi_1(p_n) = 0$ and $\pi_{i+1}(p_n) = 1 - \pi_i(p_n)$ for $1 \le i < n$ and q_n be the
point of $\{0, 1\}^n$ such that $\pi_1(q_n) = 1$ and $\pi_{i+1}(q_n) = 1 - \pi_i(q_n)$ for $1 \le i < n$.
For each positive integer n, let $D_n = \{p_n\} \times C_0$ and $E_n = \{q_n\} \times C_0$. Observe that
$C_0 = D_1 \cup D_2$ and $C_1 = E_1$. For each positive integer j, $D_{2j} = D_{2j+1} \cup D_{2j+2}$ and
$E_{2j-1} = E_{2j} \cup E_{2j+1}$. Thus, if we let $p = (0, 1, 0, 1, \dots)$ and $q = (1, 0, 1, 0, \dots)$,
then $C_0 = D_1 \cup D_3 \cup D_5 \cup \dots \cup \{p\}$ where $D_i \cap D_j = \emptyset$ if i and j are odd,

$i \neq j$. Also, $C_1 = E_2 \cup E_4 \cup E_6 \cup \cdots \cup \{q\}$ where $E_i \cap E_j = \emptyset$ if i and j are even, $i \neq j$. Furthermore, $D_2 \supseteq D_4 \supseteq D_6 \supseteq \cdots$, while $E_1 \supseteq E_3 \supseteq E_5 \supseteq \cdots$. If n is an odd positive integer, each element of $\mathcal{B}_n$ is an arc having one endpoint in D_n and the other endpoint in $E_n = E_{n+1} \cup E_{n+2}$, while if n is even, then each element of $\mathcal{B}_n$ is an arc having one endpoint in E_n and the other in $D_n = D_{n+1} \cup D_{n+2}$. Moreover, if $n \in \mathbb{N}$ and $x \in D_n \cup E_n$, then x is an endpoint of some arc in $\mathcal{B}_n$.

Choose two skew lines in three-dimensional Euclidean space and embed C_0 in one of these lines and C_1 in the other; see Fig. 2.16 where we have also shown the partitions of $C_0 = D_1 \cup D_3 \cup D_5 \cup \cdots \cup \{p\}$ and $C_1 = E_2 \cup E_4 \cup E_6 \cup \cdots \cup \{q\}$. From each point of D_1, draw a straight line interval representing an arc in $\mathcal{B}_1$ that joins it to a point of E_1, from each point of E_2, draw a straight line interval representing an arc in $\mathcal{B}_2$ that joins it to a point of D_2, and continue this process. Finally, connect the points representing p and q with a straight line interval.

A double fan. The points p and q are the endpoints of the arc A. For each n, let A_n denote the element of $\mathcal{B}_n$ having p or q as one of its endpoints. One fan is $F = A \cup (\bigcup_{i>0} A_{2i})$ with vertex q and the other is $A \cup (\bigcup_{i>0} A_{2i-1})$ having vertex p.

Triods and nonplanarity. Let v be a point of E_5 and let α be an arc in $\mathcal{B}_5$ from v to a point of D_5. Because $E_5 \subseteq E_3 \subseteq E_1$, there are arcs β and γ in $\mathcal{B}_3$ and $\mathcal{B}_1$, respectively, having v as an endpoint. Let $T_v = \alpha \cup \beta \cup \gamma$. Because D_1, D_3, and D_5 are pairwise mutually exclusive, T_v is a triod. If v and w are two different points of E_5, $T_v \cap T_w = \emptyset$. Because E_5 is uncountable, M contains uncountably many mutually exclusive triods, so M is a nonplanar continuum [8, Theorem 84, p. 222].

n-ods. To obtain a 4-od lying in M, start with a point x of E_{13}. Choose four arcs containing x, one from each of $\mathcal{B}_{13}$, $\mathcal{B}_{11}$, $\mathcal{B}_9$, and $\mathcal{B}_7$, respectively. The union of these four arcs is a 4-od. In a similar manner, we can see that for each positive integer n, there are n-ods in M for each $n \in \mathbb{N}$.

A $\sin(1/x)$-curve. There is an arc in $\mathcal{B}_1$ from the point $(0, 0, 0, \ldots)$ of D_1 to the point $(1, 0, 0, \ldots)$ of E_1. In $\mathcal{B}_2$, there is an arc from the point $(1, 0, 0, \ldots)$ of E_1 to the point $(0, 1, 0, 0, \ldots)$ of D_3. In $\mathcal{B}_3$, there is an arc from $(0, 1, 0, 0, \ldots)$ to $(1, 0, 1, 0, 0, \ldots)$ of E_3, and in $\mathcal{B}_4$, there is an arc from $(1, 0, 1, 0, 0, \ldots)$ to $(0, 1, 0, 1, 0, 0, \ldots)$ of D_5. Continuing in this way, we obtain a $\sin(1/x)$-curve having limit bar the arc A from p to q. There are other $\sin(1/x)$-curves in M. For example, instead of starting from the point with all coordinates 0, start from the point of D_1 whose coordinates are 0 except for the $4j - 1$ coordinates for $j \in \mathbb{N}$ where the coordinates are 1 and use a procedure similar to the one above. Here, the first arc would be chosen in $\mathcal{B}_1$ to a point of E_4, the second arc would be chosen from $\mathcal{B}_4$ to a point of D_5, the third arc would be chosen from $\mathcal{B}_5$ to a point of E_8, and so on. This yields a $\sin(1/x)$-curve whose intersection with the one above is the limit bar, A. $\qquad \square$

The function in Example 2.15 is conjugate to the function $g : [0, 1] \to C([0, 1])$ given by $g(t) = 1 - t$ for $0 \leq t < 1$ and $g(1) = [0, 1]$. Consequently, $\varprojlim g$ is homeomorphic to the inverse limit from Example 2.15.

Fig. 2.17 The graph of the
bonding function in
Example 2.16

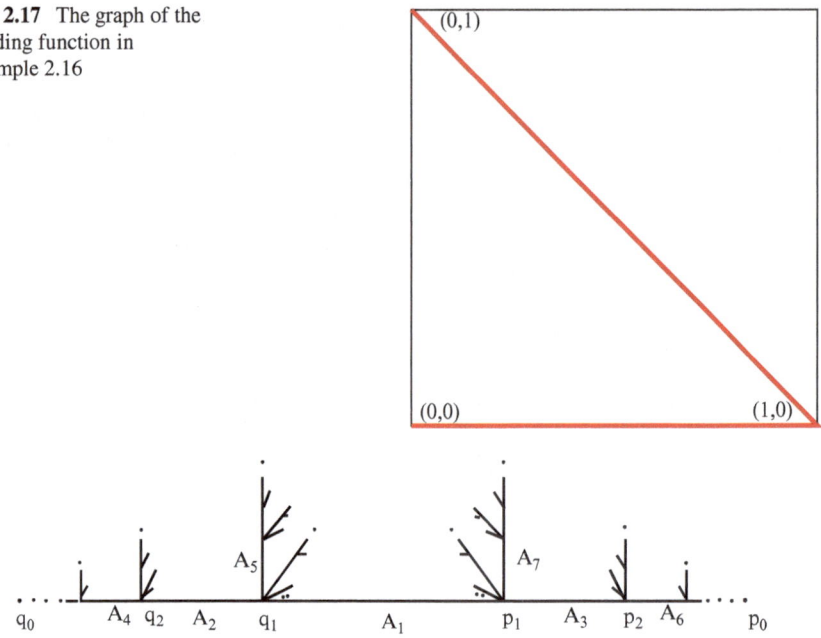

Fig. 2.18 A model of the inverse limit in Example 2.16

Example 2.16. Let $f : [0, 1] \to 2^{[0,1]}$ be given by $f(t) = \{0, 1 - t\}$ for $0 \le t \le 1$.
The inverse limit is an arcwise connected continuum that contains an arc $A = \varprojlim g$
where $g = 1 - Id$ and a Cantor set $C = \{x \in \{0, 1\}^\infty \mid$ if $x_i = 1$, then $x_{i+1} = 0\}$.
Each point of C lies in an arc that intersects A. (See Fig. 2.17 for the graph of f and
Fig. 2.18 for a model of $\varprojlim f$.)

Proof. Because $f^{-1} : [0, 1] \to C([0, 1])$, $\varprojlim f^{-1}$ is a continuum. By Theorem 2.3,
$M = \varprojlim f$ is a continuum.
Next, we show that if $p \in C$, then there is an arc containing p that intersects
A. Let p be a point of C. For each positive integer i, let α_i be an arc determined
in the following way: if $p_i = 0$, then $\alpha_i = \{x \in M \mid x_j = p_j$ for $1 \le j \le$
i and $x_{i+1} \in [0, 1], x_{i+2} = 1 - x_{i+1}, x_{i+3} = x_{i+2}, \dots\}$, while if $p_i = 1$, then
$\alpha_i = \{x \in M \mid x_j = p_j$ for $1 \le j \le i$ and $x_{i+1} = 0, x_{i+2} \in [0, 1], x_{i+3} =$
$1 - x_{i+2}, x_{i+4} = x_{i+2}, \dots\}$. Note that in the case that $p_i = 0$, $\alpha_i \cap \alpha_{i+1}$ is a single
point, while if $p_i = 1$, $\alpha_i = \alpha_{i+1}$. Because α_1 intersects A at either $(0, 1, 0, \dots)$ or
$(1, 0, 1, \dots)$, it follows that $\mathrm{Cl}(\bigcup_{i>0} \alpha_i)$ is an arc containing p and intersecting A.
That the continuum M is arcwise connected now follows.
To describe a model for the inverse limit, we proceed somewhat informally.
Because any point of the inverse limit having a 1 as a coordinate must have a 0 in
its next coordinate, let $\mathcal{S} = \{s \mid s$ is a finite sequence of 0s and 1s such that the final
term of s is 0 and if a term of s is 1, then the next term of s is 0$\}$. Then $\mathcal{S}$ is countable.

By using the two symbols 0 and 10 and writing the terms of S as strings, we may indicate an enumeration of S by $\{0, 10, 00, 010, 100, 1010, 000, 0010, 0100, 01010, 1000, 10010, 10100, 101010, \ldots\}$. Employing this enumeration of S, define a sequence of arcs in the following way. Let

$A_1 = A = \{x \in M \mid x_1 \in [0, 1]$ and $x_{i+1} = 1 - x_i$ for each positive integer $i\}$.

Using the first term of S to determine the first coordinate of all the points of an arc, let

$A_2 = \{x \in M \mid x_1 = 0, x_2 \in [0, 1]$ and $x_{i+1} = 1 - x_i$ for each positive integer $i \geq 2\}$.

Using the second term of S to determine the first two coordinates of all the points of an arc, let

$A_3 = \{x \in M \mid x_1 = 1, x_2 = 0, x_3 \in [0, 1]$ and $x_{i+1} = 1 - x_i$ for each positive integer $i \geq 3\}$.

If $n \geq 1$ and $s^n = s_1, s_2, \ldots, s_{k_n}$ is the nth term of S, let

$A_{n+1} = \{x \in M \mid x_i = s_i$ for $1 \leq i \leq k_n, x_{k_n+1} \in [0, 1]$, and $x_{i+1} = 1 - x_i$ for each positive integer $i \geq k_n + 1\}$.

Thus,

$A_4 = \{x \in M \mid x_1 = 0, x_2 = 0, x_3 \in [0, 1]$ and $x_{i+1} = 1 - x_i$ for each positive integer $i \geq 3\}$.

$A_5 = \{x \in M \mid x_1 = 0, x_2 = 1, x_3 = 0, x_4 \in [0, 1]$ and $x_{i+1} = 1 - x_i$ for each positive integer $i \geq 4\}$.

$A_6 = \{x \in M \mid x_1 = 1, x_2 = 0, x_3 = 0, x_4 \in [0, 1]$ and $x_{i+1} = 1 - x_i$ for each positive integer $i \geq 4\}$.

$A_7 = \{x \in M \mid x_1 = 1, x_2 = 0, x_3 = 1, x_4 = 0, x_5 \in [0, 1]$ and $x_{i+1} = 1 - x_i$ for each positive integer $i \geq 5\}$

.

.

.

For this sequence of arcs, $M = \mathrm{Cl}(\bigcup_{i>0} A_i)$. Let F denote the fan of Example 2.13, i.e., a fan with a sequence of arms of lengths decreasing to 0 emanating from its vertex. At each end of the arc A_1, there is a copy of the fan F attached at its vertex. At each endpoint of each copy of F, we see a copy of F again attached at its vertex, and this continues on those copies of F, et cetera. The arc A_1 has endpoints $(0, 1, 0, 1, \ldots)$ and $(1, 0, 1, 0, \ldots)$. The arcs $A_3, A_7, A_{15}, \ldots$ comprise the arms of a copy of F attached to A_1 at $(1, 0, 1, 0, \ldots)$, and the arcs $A_2, A_5, A_{11}, \ldots$ comprise the arms of a copy of F attached to A_1 at $(0, 1, 0, 1, \ldots)$. The endpoints of A_2 are $(0, 1, 0, 1, \ldots)$ and $(0, 0, 1, 0, 1, \ldots)$, and the arcs $A_4, A_9, A_{17}, \ldots$ comprise the arms of a copy of F attached to A_2 at the point $(0, 0, 1, 0, 1, \ldots)$. The endpoints of A_3 are $(1, 0, 1, 0, \ldots)$ and $(1, 0, 0, 1, 0, \ldots)$, and the arcs $A_6, A_{13}, A_{27}, \ldots$ comprise the arms of a copy of F attached to A_3 at $(1, 0, 0, 1, 0, 1, \ldots)$. See Fig. 2.18 for

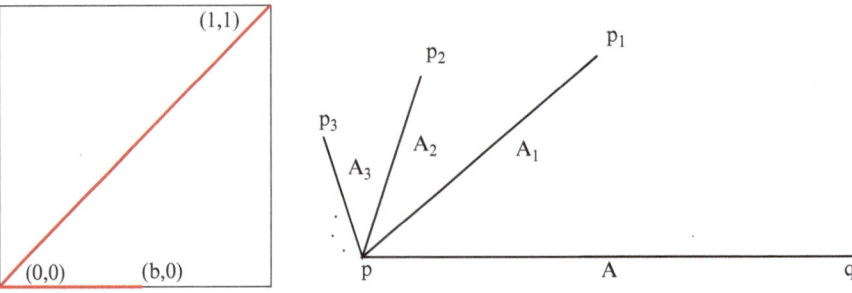

Fig. 2.19 The graph of the bonding function and a model of the inverse limit in Example 2.17

an indication of a model for M based on this informal partial description of M. In the figure, the points shown are $p_0 = (1, 0, 0, 0, \ldots)$, $p_1 = (1, 0, 1, 0, \ldots)$, $p_2 = (1, 0, 0, 1, 0, \ldots)$, $q_0 = (0, 0, 0, \ldots)$, $q_1 = (0, 1, 0, 1, \ldots)$, and $q_2 = (0, 0, 1, 0, 1, \ldots)$. $\qquad\qquad\square$

The function from Example 2.16 is conjugate to the function $g : [0, 1] \to 2^{[0,1]}$ given by $g(t) = \{1, 1 - t\}$ for $0 \le t \le 1$. Consequently, $\varprojlim g$ is homeomorphic to the inverse limit from Example 2.16. This completes our look at the collection $\mathcal{E}$.

2.8.1 Four More Similar Graphs

We now consider what happens if we use only part of the horizontal or vertical line in the construction of the eight functions. Here we only consider the four main graphs and omit reference to the four conjugate graphs. We begin with the function having a portion of the x-axis attached to the diagonal.

Example 2.17. Let b be a number such that $0 < b < 1$ and let $f : [0, 1] \to 2^{[0,1]}$ be given by $f(t) = \{0, t\}$ for $0 \le t \le b$ and $f(t) = t$ for $b < t \le 1$. Then, $\varprojlim f$ is a fan. (See Fig. 2.19 for the graph of f and a model of its inverse limit.)

Proof. Let $M = \varprojlim f$. Because $f^{-1} : [0, 1] \to C([0, 1])$, M is a continuum. Let $A = \{x \in M \mid x_1 \in [0, 1] \text{ and } x_j = x_1 \text{ for } j > 1\}$. For each positive integer i, let $A_i = \{x \in M \mid x_j = 0 \text{ for } 1 \le j \le i, x_{i+1} \in [0, b], \text{ and } x_j = x_{i+1} \text{ for } j > i + 1\}$. Then $M = A \cup (\bigcup_{i>0} A_i)$. For the purpose of identification in the model, let $p = (0, 0, 0, \ldots)$, $q = (1, 1, 1, \ldots)$, $p_1 = (0, b, b, b, \ldots)$, $p_2 = (0, 0, b, b, \ldots)$, and $p_3 = (0, 0, 0, b, b, \ldots)$. $\qquad\qquad\square$

Example 2.18. Let b be a number such that $0 < b < 1$ and let $f : [0, 1] \to C([0, 1])$ be given by $f(t) = t$ for $0 \le t < 1$ and $f(1) = [b, 1]$. Then, $\varprojlim f$ is a fan. (See Fig. 2.20 for the graph of f and a model of its inverse limit.)

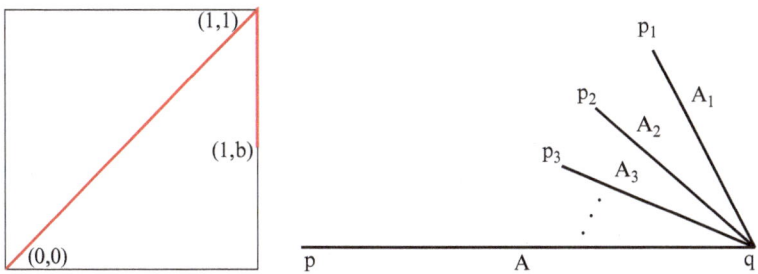

Fig. 2.20 The graph of the bonding function and a model of the inverse limit in Example 2.18

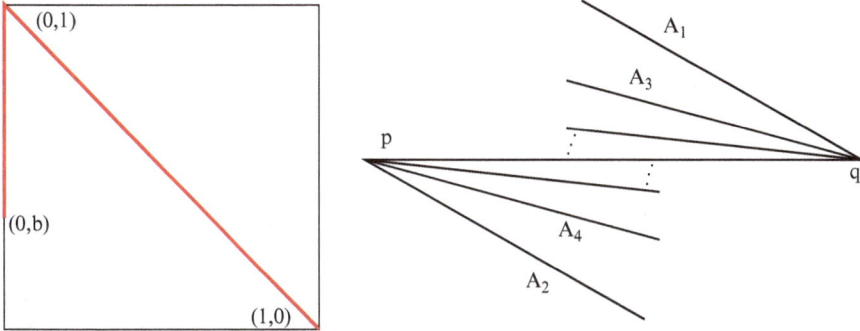

Fig. 2.21 The graph of the bonding function and a model of the inverse limit in Example 2.19

Proof. Let $M = \varprojlim f$. Because $f : [0, 1] \to C([0, 1])$, M is a continuum. Let $A = \{x \in M \mid x_1 \in [0, 1]$ and $x_j = x_1$ for $j > 1\}$. For each positive integer i, let $A_i = \{x \in M \mid x_1 \in [b, 1], x_j = x_1$ for $1 \leq j \leq i$, and $x_j = 1$ for $j > i\}$. Then $M = A \cup (\bigcup_{i>0} A_i)$. For the purpose of identification in the model, let $p = (0, 0, 0, \dots)$, $q = (1, 1, 1, \dots)$, $p_1 = (b, 1, 1, 1, \dots)$, $p_2 = (b, b, 1, 1, \dots)$, and $p_3 = (b, b, b, 1, 1, \dots)$. □

Example 2.19. Let b be a number such that $0 < b < 1$ and let $f : [0, 1] \to C([0, 1])$ be given by $f(0) = [b, 1]$ and $f(t) = 1 - t$ for $0 < t \leq 1$. Then $\varprojlim f$ is the union of two fans that intersect in an arc. (See Fig. 2.21 for the graph of f and a model of its inverse limit.)

Proof. Let $M = \varprojlim f$. Because $f : [0, 1] \to C([0, 1])$, M is a continuum. Let $A = \{x \in M \mid x_1 \in [0, 1]$ and $x_{j+1} = 1 - x_j$ for $j \geq 1\}$. If i is a positive integer, let $A_{2i-1} = \{x \in M \mid x_1 \in [b, 1], x_{2i} = 0,$ and $x_{j+1} = 1 - x_j$ for $j \neq 2i - 1$ and $j \geq 1\}$ and $A_{2i} = \{x \in M \mid x_1 \in [b, 1], x_{2i+1} = 0$ and $x_{j+1} = 1 - x_j$ for $j \neq 2i$ and $j \geq 1\}$. Then $M = A \cup (\bigcup_{i>0} A_i)$. In the model, the point $p = (0, 1, 0, 1, \dots)$, and the point $q = (1, 0, 1, 0, \dots)$. This inverse limit is homeomorphic to the union of two copies of the inverse limit from Example 2.18 having the arc A in common. In the model, the point p is $(0, 1, 0, 1, \dots)$ and $q = (1, 0, 1, 0, \dots)$. □

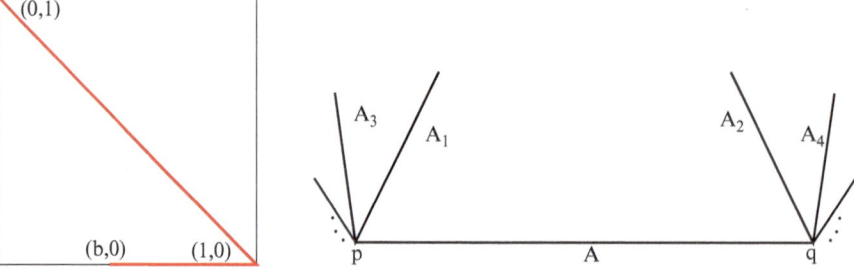

Fig. 2.22 The graph of the bonding function and a model of the inverse limit in Example 2.20

Example 2.20. Let b be a number such that $0 < b < 1$ and let $f : [0, 1] \to 2^{[0,1]}$ be given by $f(t) = 1 - t$ for $0 \le t < b$ and $f(t) = \{0, 1 - t\}$ for $b \le t \le 1$. Then $\varprojlim f$ is the union of two fans. (See Fig. 2.22 for the graph of f and a model of its inverse limit.)

Proof. Let $M = \varprojlim f$. Because $f^{-1} : [0, 1] \to C([0, 1])$, M is a continuum. Let $A = \{x \in M \mid x_1 \in [0, 1]$ and $x_{j+1} = 1 - x_j$ for $j \ge 1\}$. For each positive integer i, let $A_{2i} = \{x \in M \mid x_1 = 0, x_{2i} \in [0, b]$, and if $i \ne 2i$, then $x_{j+1} = 1 - x_j$ for $j \ge 1\}$ and $A_{2i+1} = \{x \in M \mid x_1 = 1, x_{2i+1} \in [0, b]$, and if $i \ne 2i + 1$, then $x_{j+1} = 1 - x_j$ for $j \ge 1\}$. This inverse limit is homeomorphic to the union of two copies of the inverse limit from Example 2.17 having the arc A in common. In the model, the point p is $(0, 1, 0, 1, \ldots)$, and $q = (1, 0, 1, 0, \ldots)$ (Fig. 2.22). $\square$

2.9   Additional Examples

We examine some additional examples in this section. Our first function yields a fan as its inverse limit. In fact, the inverse limit in Example 2.21 is homeomorphic to the inverse limit in Example 2.14.

Example 2.21. Let $f : [0, 1] \to C([0, 1])$ be given by $f(t) = t$ for $0 \le t < 1/2$, $f(1/2) = [1/2, 1]$, and $f(t) = 1 - t$ for $1/2 < t \le 1$. Then, $\varprojlim f$ is a fan with vertex $v = (1/2, 1/2, 1/2, \ldots)$. (See Fig. 2.23 for the graph of f and a model of its inverse limit.)

Proof. Let $M = \varprojlim f$ and let $B_0 = \{x \in M \mid x_1 \in [0, 1/2]$ and $x_j = x_1$ for $j > 1\}$. Let i be a positive integer and let $B_i = \{x \in M \mid x_i \in [1/2, 1]\}$. Note that if $x \in B_i$, then $x_j = 1/2$ for $j > i$, and if $i > 1$, then $x_j \in [0, 1/2]$ for $j < i$. Further, if $i > 2$ and $x \in B_i$, then $x_i = 1 - x_{i-1}$ and $x_j = x_1$ for $j \le i - 1$. There is a natural homeomorphism from $[1/2, 1]$ onto B_i, so B_i is an arc and $v = (1/2, 1/2, 1/2, \ldots) \in B_i$ for each $i \ge 0$. Moreover, $M = \bigcup_{i \ge 0} B_i$.

We now show that M is homeomorphic to the inverse limit from Example 2.14. Denote by N that inverse limit. Then $N = \bigcup_{i \ge 0} A_i$ where $A_0 = \{x \in N \mid x_1 \in [0, 1]$

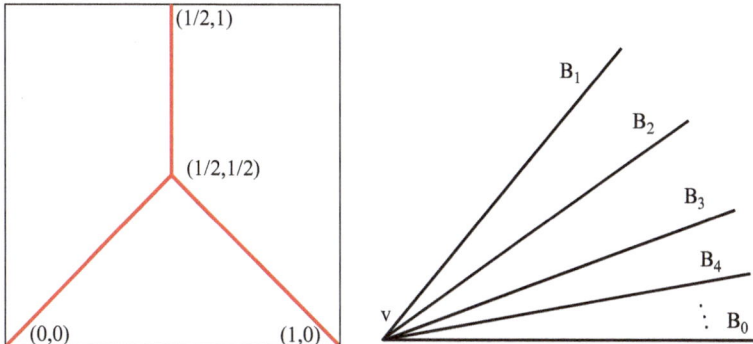

Fig. 2.23 The graph of the bonding function and a model of the inverse limit in Example 2.21

Fig. 2.24 The graph of the bonding function in Example 2.22

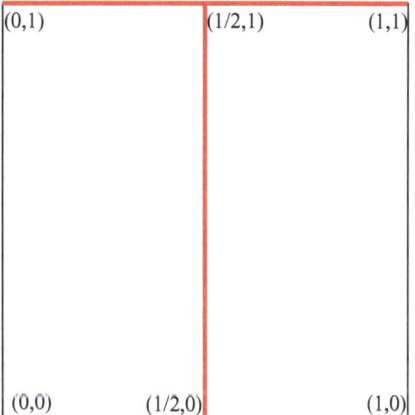

and $x_i = x_1$ for $i \geq 1$}, while $A_i = \{x \in N \mid x_1 \in [0,1], x_j = x_1$ for $1 \leq j \leq i$, and $x_j = 1$ for $j > i\}$ for $i \geq 1$. If $x \in B_0$, let $h(x) = (2x_1, 2x_1, 2x_1, \dots)$, and for $i \geq 1$ and $x \in B_i$, let $h(x)$ be the element of A_i having ith coordinate $2x_i - 1$. Then, h is a homeomorphism from $\bigcup_{i \geq 0} B_i$ onto $\bigcup_{i \geq 0} A_i$, and we have that M is homeomorphic to the inverse limit from Example 2.14. □

We describe two additional examples. These examples were worked out by students in Mexico during the two-week short course on which much of the material in this book is based. Both examples grew out of an assignment in which they were to choose an embedding of a letter of the alphabet into $[0,1]^2$ so that it forms the graph of an upper semicontinuous function and then determine its inverse limit (see Problem 6.61). The inverse limit in Example 2.22 is a familiar dendroid.

Example 2.22 (A dendroid with a Cantor set of endpoints). Let $f : [0,1] \to 2^{[0,1]}$ be given by $f(t) = \{0,1\}$ for $t \neq 1/2$ and $f(1/2) = [0,1]$. Then, $\varprojlim f$ is a dendroid having a Cantor set of endpoints. (See Fig. 2.24 for the graph of f and Fig. 2.25 for a model of the inverse limit.)

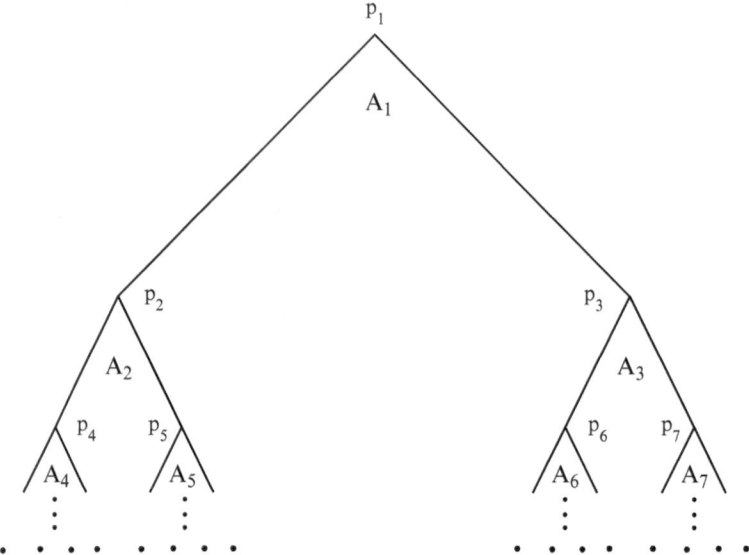

Fig. 2.25 A model of the inverse limit in Example 2.22

Proof. Let $M = \varprojlim f$, a continuum because $f^{-1} : [0,1] \to C([0,1])$. Let $A_1 = \{x \in M \mid x_i = 1/2 \text{ for } i > 1\}$, $A_2 = \{x \in M \mid x_i = 1/2 \text{ for } i > 2 \text{ and } x_1 = 0\}$, and $A_3 = \{x \in M \mid x_i = 1/2 \text{ for } i > 2 \text{ and } x_1 = 1\}$. In general, suppose $n \in \mathbb{N}$. There is a positive integer k such that the base 2 representation of n is $a_k a_{k-1} \cdots a_0$ with $a_k = 1$. For $n \geq 2$, let $A_n = \{x \in M \mid x_{k+1} \in [0,1], x_i = 1/2 \text{ for } i > k+1 \text{ and } x_i = a_{k-i} \text{ for } 1 \leq i \leq k\}$. Note that $A_1 \cap A_2 = \{(0, 1/2, 1/2, 1/2, \dots)\}$ and $A_1 \cap A_3 = \{(1, 1/2, 1/2, 1/2, \dots)\}$. In fact, letting $p_1 = (1/2, 1/2, 1/2, \dots)$ and, for $n > 1$, $p_n = (a_{k-1}, a_{k-2}, \dots, a_0, 1/2, 1/2, \dots)$, then $A_1 \cap A_2 = \{p_2\}$, $A_1 \cap A_3 = \{p_3\}$, $A_2 \cap A_4 = \{p_4\}$, etc. We see that p_2 is an endpoint of A_1 and an interior point of A_2. In general, A_n is an arc with endpoints p_{2n} and p_{2n+1}; the point p_n is an interior point of A_n. Note that $M = \text{Cl}(\bigcup_{i>0} A_i)$ (where Cl denotes the closure). $\qquad \square$

Example 2.23. Let $f : [0,1] \to C[0,1])$ be given by $f(0) = f(1) = [0,1]$ and $f(t) = \{1/2\}$ for $t \notin \{0,1\}$. The inverse limit is a continuum that is the union of a sequence $B_0, B_1, B_2, \dots$ of compacta such that $B_{i+1} \cap B_i$ is a Cantor set, B_0 is a single point, and, for each positive integer i, B_i is homeomorphic to a product of a Cantor set and an arc (Figs. 2.26 and 2.27).

Proof. Let $M = \varprojlim f$, a continuum by Theorem 2.7. Let $B_0 = \{(1/2, 1/2, 1/2, \dots)\}$, let $B_1 = \{x \in M \mid x_1 \in [0,1] \text{ and } x_j \in \{0,1\} \text{ for } j > 1\}$, and, for each integer $i > 1$, let $B_i = \{x \in M \mid x_j = 1/2 \text{ for } j < i, x_i \in [0,1], \text{ and } x_j \in \{0,1\} \text{ for } j > i\}$. Note that $B_i \cap B_{i+1} = \{x \in M \mid x_j = 1/2 \text{ for } j \leq i \text{ and } x_j \in \{0,1\} \text{ for } j > i\}$. Note that $M = \bigcup_{i \geq 0} B_i$. Because each arc component of B_{i+1} intersects two arc components of B_i and both points of intersection are

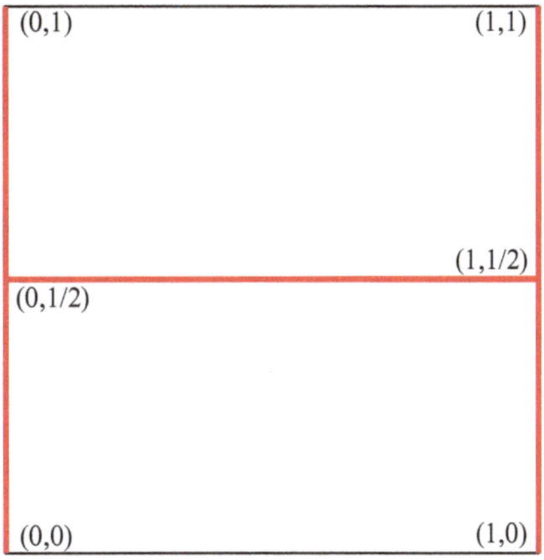

Fig. 2.26 The graph of the
bonding function in
Example 2.23

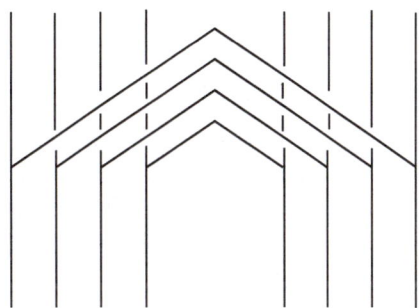

Fig. 2.27 A model of
$B_1 \cup B_2$ in Example 2.23, a
set homeomorphic to
$B_i \cup B_{i+1}$ for each positive
integer i

interior to its respective arc component of B_i, M contains uncountably many triods.
Consequently, M is a nonplanar continuum, and a model for M is not so simple to
depict. In Fig. 2.27, we indicate a typical set $B_i \cup B_{i+1}$ for $i \geq 1$.   □

There is a connection between Examples 2.23 and 2.22. The components of $B_1 \cup$
B_2 from Example 2.23 are homeomorphic to $A_1 \cup A_2 \cup A_3$ from Example 2.22
(there are uncountably many of these components in Example 2.23), whereas the
components of $B_1 \cup B_2 \cup B_3$ are homeomorphic to $A_1 \cup A_2 \cup \cdots A_7$ (see Fig. 2.25).

2.10 Nonconnected Inverse Limits

In [10, Example 3.4], Nall provides a simple example of a function with a
connected graph whose inverse limit is not connected. His proof involves the use
of Theorem 2.2. This is our next example.

Example 2.24 (Nall; An inverse that is not connected). Let $f : [0, 1] \to 2^{[0,1]}$ be
given by $f(t) = t/2$ for $0 \leq t < 1/2$ and $f(t) = \{t/2, 2t - 1\}$ for $1/2 \leq t \leq 1$.
Then, $G(f)$ is connected, but $\varprojlim f$ is not connected. (See Fig. 2.28 for the graphs
of f and f^2.)

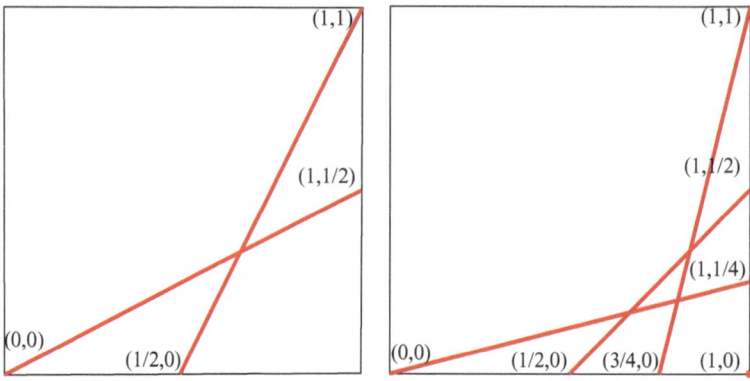

Fig. 2.28 Graphs of the functions f and f^2 in Example 2.24

Proof. The graph of f is easily seen to be connected. The graph of f^2 contains the point $(1, 0)$ as an isolated point.   □

There exist functions for which it is quite difficult to use Theorem 2.2 to determine that the inverse limit is not connected. Our next example provides a sequence of such functions. In this sequence, as n increases, so does the difficulty of making use of Theorem 2.2. Our proof is the same as the one provided in [4]. We begin with a lemma.

Lemma 2.3. *Suppose* $f : [0, 1] \rightarrow 2^{[0,1]}$ *is an upper semicontinuous function. If k is a positive integer, then* $G(f^{k+1}) = \{(x, y) \in [0, 1]^2 \mid$ *there exists a point $t \in [0, 1]$ such that $x \in f^{-1}(t)$ and $y \in f^k(t)\}$.*

Proof. $y \in f^{k+1}(x)$ if and only if there is a point $t \in [0, 1]$ such that $t \in f(x)$ and $y \in f^k(t)$ therefore, we have that $y \in f^{k+1}(x)$ if and only if there is a point $t \in [0, 1]$ such that $x \in f^{-1}(t)$ and $y \in f^k(t)$.   □

Example 2.25. Let n be an integer greater than 1. Let $f_n : [0, 1] \rightarrow 2^{[0,1]}$ be given by $f_n(t) = t$ for $0 \leq t < 1/n$, $f_n(t) = \{t, 2t - 2/n, t - 1/n\}$ for $1/n \leq t \leq 2/n$, and $f_n(t) = \{t, t - 1/n\}$ for $2/n < t \leq 1$. Then, for $1 \leq k < n$, $G(f_n^k)$ is connected, but $G(f_n^n)$ is not connected (Fig. 2.29).

Proof. Choose a positive integer $n \geq 2$. Observe that f_n is the union of three homeomorphisms:

$g_1 = Id_{[0,1]}$,
$g_2 : [1/n, 1] \rightarrow [0, 1-1/n]$ where $g_2(x) = x - 1/n$,
$g_3 : [1/n, 2/n] \rightarrow [0, 2/n]$ where $g_3(x) = 2x - 2/n$.

It is clear that $G(f_n)$ is connected because $G(g_3)$ intersects both $G(g_1)$ and $G(g_2)$. Note that the points $(0, 0)$ and $(1/n, 0)$ belong to $G(f_n)$, and the entire graph of $G(f_n)$ lies in $[0, 1-1/n]^2$ except for two nonseparating half-open intervals

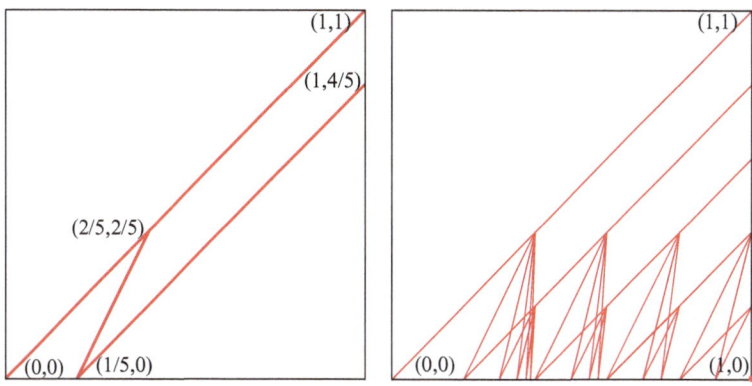

Fig. 2.29 Graphs of the functions f_5 and f_5^5 in Example 2.25

lying in the strip $(1-1/n, 1] \times [0, 1]$. Thus, $G(f_n | [0, 1-1/n])$ is connected. Clearly, $G(f_n | [0, 2/n])$ is connected, whereas $f_n([0, 2/n]) = [0, 2/n]$ and $f_n([0, 1-1/n]) = [0, 1-1/n]$.

Let $\varphi_1 : [0, 1]^2 \to [0, 1]^2$ be given by $\varphi_1(x, y) = (x, y)$, let $\varphi_2 : [0, 1-1/n]^2 \to [1/n, 1] \times [0, 1-1/n]$ be given by $\varphi_2(x, y) = (x + 1/n, y)$, and let $\varphi_3 : [0, 2/n]^2 \to [1/n, 2/n] \times [0, 2/n]$ be given by $\varphi_3(x, y) = (x/2 + 1/n, y)$. Note that $\varphi_1(x, y) = (g_1^{-1}(x), y)$ for $(x, y) \in [0, 1]^2$; $\varphi_2(x, y) = (g_2^{-1}(x), y)$ for $(x, y) \in [0, 1-1/n]^2$ and $\varphi_3(x, y) = (g_3^{-1}(x), y)$ for $(x, y) \in [0, 2/n]^2$.

We now show that if $1 \leq k \leq n - 1$, then $G(f_n^{k+1}) = \varphi_1(G(f_n^k)) \cup \varphi_2(G(f_n^k | [0, 1-1/n]) \cup \varphi_3(G(f_n^k | [0, 2/n]) - \{(2/n, 0)\})$. To see this first, let (x, y) be a point of $G(f_n^{k+1})$. By Lemma 2.3, there is a point $t \in [0, 1]$ such that $x \in f_n^{-1}(t)$ and $y \in f_n^k(t)$. There is an integer i, $1 \leq i \leq 3$, such that $x = g_i^{-1}(t)$, and, for such an i, $(x, y) = \varphi_i(t, y)$ with $(t, y) \in G(f_n^k)$. If $i = 1$, $(x, y) \in \varphi_1(G(f_n^k))$. If $i = 2$, then $0 \leq t \leq 1-1/n$, so $(x, y) \in \varphi_2(G(f_n^k | [0, 1-1/n]))$. If $i = 3$ and $(x, y) \neq (2/n, 0)$, then $t \in [0, 2/n]$ and $(x, y) \in \varphi_3(G(f_n^k | [0, 2/n]) - \{(2/n, 0)\})$. In case $(x, y) = (2/n, 0)$, $(x, y) = \varphi_2(1/n, 0)$, so $(x, y) \in \varphi_2(G(f_n^k | [0, 1-1/n]))$. On the other hand, if $(x, y) \in \varphi_1(G(f_n^k)) \cup \varphi_2(G(f_n^k | [0, 1-1/n]) \cup \varphi_3(G(f_n^k | [0, 2/n]) - \{(2/n, 0)\})$, then for some $i, 1 \leq i \leq 3$ and some point $t \in [0, 1]$, $x \in g_i^{-1}(t)$, and $y \in f_n^k(t)$. It follows from Lemma 2.3 that $(x, y) \in G(f_n^{k+1})$.

Next, we proceed inductively to show that $G(f_n^k)$ is a connected set containing $(0, 0)$ and $(m/n, 0)$ for $1 \leq k \leq n - 1$ and $1 \leq m \leq k$. We have observed this to be true for $k = 1$ because $G(f_n)$ is connected as are $G(f_n | [0, 1-1/n])$ and $G(f_n | [0, 2/n])$ and $(0, 0)$ and $(1/n, 0)$ are points of $G(f_n)$.

Suppose j is an integer, $1 \leq j < n - 1$, such that $G(f_n^j)$ is a connected set as are $G(f_n^j | [0, 1-1/n])$ and $G(f_n^j | [0, 2/n] - \{(2/n, 0)\})$ (we only need to remove the point $(2/n, 0)$ when $j > 1$ because, of course, this point is not in $G(f_n)$). Suppose also that $(0, 0)$ and $(m/n, 0)$ are in $G(f_n^j)$ for $1 \leq m \leq j$. Then, $\varphi_1(G(f_n^j))$ is connected as are $\varphi_2(G(f_n^j | [0, 1-1/n]))$ and $\varphi_3(G(f_n^j | [0, 2/n] - \{(2/n, 0)\})$.

Fig. 2.30 The graph of the bonding function in Example 2.26

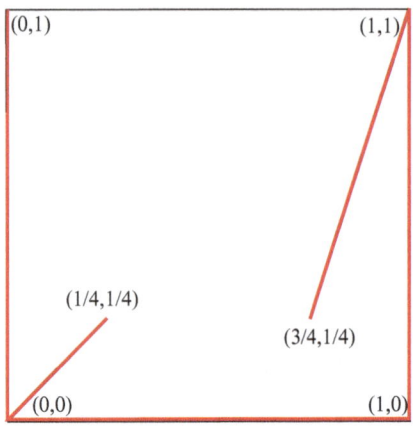

The point $(1/n, 0)$ belongs to all three of these sets because $\varphi_1(1/n, 0) = (1/n, 0)$ and $(1/n, 0) \in G(f_n^j)$, whereas $\varphi_2(0, 0) = \varphi_3(0, 0) = (1/n, 0)$ and $(0, 0)$ belongs to both $G(f_n^j |[0, 1-1/n])$ and $G(f_n^j |[0, 2/n])$. Thus, $G(f_n^{j+1})$ is connected and contains $(0, 0)$ because $\varphi_1(0, 0) = (0, 0)$. Further, the entire graph of f_n^{j+1} lies in $[0, 1-1/n]^2$ except for $j + 2$ nonseparating half-open intervals lying in the strip $(1-1/n, 1] \times [0, 1]$ (the extra one that is not part of the graph of f_n^j comes from $\varphi_2(G(f_n^j |[0, 1-1/n])))$, so $G(f_n^{j+1} |[0, 1-1/n])$ is connected. Finally, $G(f_n^{j+1} |[0, 2/n]) - \{(2/n, 0)\}$ is connected. To see this, observe that the portion of $G(f_n^j)$ mapped into $[0, 2/n]^2$ by φ_2 is the union of the straight line interval from $(0, 0)$ to $(1/n, 1/n)$ and the single point $(1/n, 0)$. Thus, $\varphi_2(G(f_n^j |[0, 1-1/n])) \cap [0, 2/n]^2$ is the union of the straight line interval from $(1/n, 0)$ to $(2/n, 1/n)$ and the point $(2/n, 0)$. It follows that $G(f_n^{j+1} |[0, 2/n]) - \{(2/n, 0)\}$ is connected being the union of three connected sets $\varphi_1(G(f_n^j |[0, 2/n]) - \{(2/n, 0)\}$, the straight line interval from $(1/n, 0)$ to $(2/n, 1/n)$, and $\varphi_3(G(f_n^j |[0, 2/n] - \{(2/n, 0)\}))$, all containing $(1/n, 0)$. Because $(m/n, 0)$ is in $G(f_n^j)$ for $1 \le m \le j$ and $\varphi_2(i/n, 0) = ((i + 1)/n, 0)$ for each i, $1 \le i \le j$, $(m/n, 0) \in G(f_n^{j+1})$ for $1 \le m \le j + 1$.

Therefore, we have that $G(f_n^k)$ is connected for $1 \le k \le n-1$ and $(1-1/n, 0) \in G(f_n^{n-1})$. It now follows that $\varphi_2(1-1/n, 0) = (1, 0)$ is in $G(f_n^n)$. However, $(1, 0)$ is an isolated point of $G(f_n^n)$. To see this, observe that $f_n^n(1)$ is a discrete set with minimum 0 and $f_n^{-n}(0)$ is a discrete set with maximum 1. Because $G(f_n^n)$ has an isolated point, it is not connected. $\qquad \square$

One cannot always rely on Theorem 2.2 to detect that an inverse limit is not connected. The following example due to Jonathan Meddaugh demonstrates this. Example 2.26 is a modification of [5, Example 1, p. 266].

Example 2.26 (Meddaugh). Let $f : [0, 1] \to 2^{[0,1]}$ be given by $f(0) = [0, 1]$, $f(t) = \{t, 0\}$ for $0 < t \le 1/4$, $f(t) = 0$ for $1/4 < t < 3/4$, $f(t) = \{3t - 2, 0\}$ for $3/4 \le t < 1$, and $f(1) = [0, 1]$. Then $G(f)$ is connected and $G(f^n) = [0, 1]^2$ for $n > 1$, but $\varprojlim f$ is not connected. (See Fig. 2.30 for the graph of f.)

Fig. 2.31 The graph of the bonding function in Example 2.27

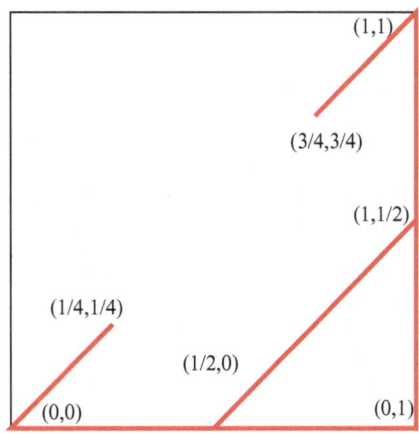

Proof. It is clear that $G(f)$ is connected and $G(f^n) = [0,1]^2$ for $n > 1$. Let $M = \varprojlim f$. The set $N = \{x \in M \mid x_1 = x_2 = 1/4,$ and $x_3 = 3/4\}$ is both open and closed in M because N is closed and $N = ((1/8, 3/8) \times (1/8, 3/8) \times (5/8, 7/8) \times Q) \cap M$. Thus, M is not connected. $\qquad\square$

A second example of an inverse limit that is not connected but the graphs of all composites of the bonding functions are connected is an example in a recent paper by Greenwood and Kennedy in which they showed that the inverse limit is not connected [1, Example 1.4, p. 58].

Example 2.27. Let $f : [0, 1] \to 2^{[0,1]}$ be given by $f(t) = \{0, t\}$ for $0 \le t \le 1/4$, $f(t) = 0$ for $1/4 < t < 1/2$, $f(t) = \{t - 1/2, 0\}$ for $1/2 < t < 3/4$, $f(t) = \{t, t - 1/2, 0\}$ for $3/4 \le t < 1$, and $f(1) = [0, 1]$. Then, $G(f)$ is connected and $f^n = f$ for each $n \in \mathbb{N}$, but $\varprojlim f$ is not connected. (See Fig. 2.31 for the graph of f.)

Proof. It is clear that $G(f)$ is connected. It is not difficult to show that $f^2 = f$, and therefore, $f^n = f$ for each $n \in \mathbb{N}$. Let $M = \varprojlim f$ and let $N = \{x \in M \mid x_1 = x_2 = 1/4$ and $x_3 = x_4 = 3/4\}$; N is a closed subset of M. Let U be the basic open set $(1/8, 3/8) \times (1/8, 3/8) \times (5/8, 7/8) \times (5/8, 7/8) \times Q$. Note that $N \subseteq U \cap M$. To see that $N = U \cap M$, suppose $x \in U \cap M$. Because $x_1 \in (1/8, 3/8)$ and $x_2 \in (1/8, 3/8)$, we observe that $x_2 \in (1/8, 1/4]$. From $x_2 \in (1/8, 1/4]$ and $x_3 \in (5/8, 7/8)$, it follows that $x_3 \in (5/8, 3/4]$. However, with $x_3 \in (5/8, 3/4]$ and $x_4 \in (5/8, 7/8)$, it follows that $x_4 = 3/4$. Because $x_4 = 3/4$ and $x_3 \in (5/8, 7/8)$, $x_3 = 3/4$; $x_3 = 3/4$ and $x_2 \in (1/8, 3/8)$ yields $x_2 = 1/4$. Because $x_2 = 1/4$ and $x_1 \in (1/8, 3/8)$, $x_1 = 1/4$. Thus, $x \in N \cap M$. Because M contains a closed set N that is open in M, M is not connected. $\qquad\square$

References

1. Greenwood, S., Kennedy, J.: Connected generalized inverse limits. Topology Appl. **159**, 57–68 (2012)
2. Hurewicz, W.: Über oberhalb-stetige Zerlegungen von Punktmengen in Kontinua. Fund. Math. **15**, 57–60 (1930)
3. Ingram, W.T.: Inverse limits with upper semi-continuous bonding functions: Problems and partial solutions. Topology Proc. **36**, 353–373 (2010)
4. Ingram, W.T.: Concerning nonconnected inverse limits with upper semi-continuous set-valued functions. Topology Proc. **40**, 203–214 (2012)
5. Ingram, W.T., Mahavier, W.S.: Inverse limits of upper semi-continuous set valued functions. Houston J. Math. **32**, 119–130 (2006)
6. Ingram, W.T., Mahavier, W.S.: Inverse limits: From continua to Chaos. In: Developments in Mathematics, vol. 25. Springer, New York (2012)
7. Mahavier, W.S.: Inverse limits with subsets of $[0, 1] \times [0, 1]$. Topology Appl. **141**, 225–231 (2004)
8. Moore, R.L.: Foundations of Point Set Theory, vol. 13, rev. edn. American Mathematical Society Colloquium Publications, Providence (1962)
9. Nadler, S.B., Jr.: Continuum Theory. Marcel-Dekker, New York (1992)
10. Nall, V.: Connected inverse limits with a set-valued function. Topology Proc. **40**, 167–177 (2012)

Chapter 3
Mappings versus Set-Valued Functions

Abstract Inverse limits with upper semicontinuous bonding functions exhibit fundamental differences from inverse limits with mappings in the sense that the theorems that hold when the bonding functions in an inverse limit sequence are mappings almost always fail if the bonding functions are set-valued. This chapter is devoted to examining some of those differences. Of course, these differences provide a source for research questions.

3.1 Introduction

Most of the tools available when the bonding functions in an inverse limit sequence are mappings fail when the bonding functions are set-valued. In this chapter, we examine a number of these tools. However, rather than viewing these failures as being negative, instead we see it as an opportunity for additional research into inverse limits with set-valued functions.

There are some theorems that we have seen that do carry over from the setting of inverse limits with mappings to inverse limits with set-valued functions. These include the fundamental existence theorem, Theorem 1.6, and the topological conjugacy theorem, Theorem 2.9. Beyond these results, one can normally expect to need additional hypotheses or special circumstances in order for theorems on inverse limits with mappings hold for inverse limits with set-valued functions.

3.2 The Subsequence Theorem

One valuable tool in inverse limits with mappings is the subsequence theorem. For sequences of mappings of $[0, 1]$, its statement is found below. Its proof is not difficult. A proof in a general setting can be found in [4, Sect. 2.10.5, p. 119]. In the statement of the theorem, we employ the convention that f_{ij} denotes $f_i \circ f_{i+1} \circ \cdots \circ f_{j-1}$.

W.T. Ingram, *An Introduction to Inverse Limits with Set-valued Functions*,
SpringerBriefs in Mathematics, DOI 10.1007/978-1-4614-4487-9_3,
© W.T. Ingram 2012

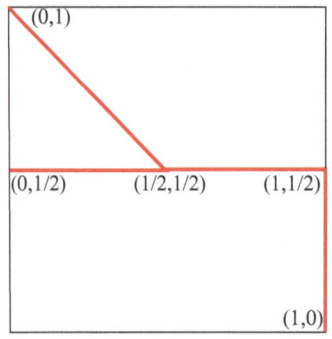 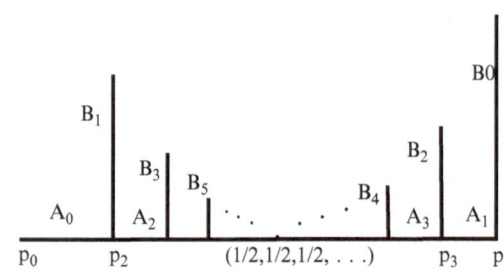

Fig. 3.1 The graph of the bonding function and a model of the inverse limit in Example 3.1

Theorem 3.1 (The subsequence theorem). *If f is a sequence of mappings of $[0, 1]$ into $[0, 1]$ and n is an increasing sequence of positive integers, then $\varprojlim f$ is homeomorphic to $\varprojlim g$, where, for each positive integer i, $g_i = f_{n_i\, n_{i+1}}$.*

This theorem fails for inverse limits with set-valued functions. We provide a couple of examples from [3].

Example 3.1 ($\varprojlim f$ and $\varprojlim f^2$ may not be homeomorphic). Let $f : [0, 1] \to 2^{[0,1]}$ be the upper semicontinuous function given by $f(t) = \{1-t, 1/2\}$ for $0 \le t \le 1/2$, $f(t) = 1/2$ for $1/2 < t < 1$, and $f(1) = [0, 1/2]$. Then, $\varprojlim f$ contains a triod, but f^2 is the function from Example 2.6 ($f^2(0) = [0, 1/2]$, $f^2(t) = 1/2$ for $0 < t < 1$, and $f(1) = [1/2, 1]$) so $\varprojlim f^2$ is an arc (see Fig. 3.1 for the graph of f and its inverse limit; see Fig. 2.5 for the graph of f^2).

Proof. Let $M = \varprojlim f$. Let $\alpha = \{x \in M \mid x_1 \in [1/2, 1], x_2 = 1 - x_1$, and $x_{2j-1} = 1$, while $x_{2j} = 0$ for each integer $j > 1\}$. Let $\beta = \{x \in M \mid x_1 = x_2 = 1/2, x_3 \in [1/2, 1], x_4 = 1 - x_3$ and $x_{2j-1} = 1$, while $x_{2j} = 0$ for each integer $j > 2\}$. Let $\gamma = \{x \in M \mid x_1 = 1/2, x_2 \in [0, 1/2]$, and $x_{2j-1} = 1$, while $x_{2j} = 0$ for each integer $j > 1\}$. Then, $(1/2, 1/2, 1, 0, 1, 0, \ldots)$ is the only point common to any two of the arcs α, β, and γ, so $\alpha \cup \beta \cup \gamma$ is a triod. It is not difficult to see that f^2 is the bonding function from Example 2.6, so $\varprojlim f^2$ is an arc. (In Fig. 3.1, $A_0 = \alpha$, $A_2 = \beta$, and $B_1 = \gamma$.)

We conclude the discussion of Example 3.1 with a description of our model of $M = \varprojlim f$ as shown in Fig. 3.1. Let $A_0 = \{x \in M \mid x_1 \in [1/2, 1], x_2 = 1 - x_1$, and $x_{2j-1} = 1$, while $x_{2j} = 0$ for each integer $j > 1\}$ (i.e., A_0 is the arc α described in the previous paragraph). For each positive integer i, let $A_i = \{x \in M \mid x_j = 1/2$ for $1 \le j \le i, x_{i+1} \in [1/2, 1], x_{i+2} = 1 - x_i, x_{i+j} = 1$, for each odd integer $j \ge 3$, while $x_{i+j} = 0$, for each even integer $j \ge 4\}$. Let $B_0 = \{x \in M \mid x_1 \in [0, 1/2], x_{2j} = 1$, while $x_{2j+1} = 0$ for $j \ge 1\}$ and, for each positive integer i, let $B_i = \{x \in M \mid x_j = 1/2$ for $1 \le j \le i, x_{i+1} \in [0, 1/2]$ and $x_{i+j} = 1$ for each even positive integer j, while $x_{i+j} = 0$ for each odd positive integer $j \geq 3$}. (Here, A_2 is β and B_1 is γ as described above.) Then, $M = (\bigcup_{i \geq 0} A_i) \cup (\bigcup_{i \geq 0} B_i) \cup \{(1/2, 1/2, 1/2, \ldots)\}$. Let $p_0 = (1, 0, 1, 0, \ldots)$ and, for each positive integer i, let p_i be the point of M such that $\pi_j(p_i) = 1/2$ for $1 \leq i \leq j$ and $\pi_{i+j}(p_i) = 1$ for each odd positive integer j and $\pi_{i+j}(p_i) = 0$ for each even positive integer j. Note that $A_i \cap A_{i+2} = \{p_{i+2}\}$ for $i \geq 0$, $B_0 \cap A_1 = \{p_1\}$ and $B_i \cap A_{i-1} \cap A_{i+1} = \{p_{i+1}\}$ for each positive integer i.

The inverse limit in Example 3.1 is homeomorphic to the union of two copies of the inverse limit from Example 2.4 that intersect at a single point. □

Our next example is one we have already discussed, the Hurewicz continuum of Example 2.11. There, we were interested in the fact that the inverse limit is a continuum even though the bonding function does not have connected values. Here we show that the subsequence theorem fails for this inverse limit sequence.

Example 3.2. Let $g_1 : [0, 1] \to [0, 1]$ be the mapping given by $g_1(t) = t + 1/2$ for $0 \leq t \leq 1/2$ and $g_1(t) = 3/2 - t$ for $1/2 \leq t \leq 1$. Let $g_2 : [0, 1] \to [0, 1]$ be the mapping given by $g_2(t) = 1/2 - t$ for $0 \leq t \leq 1/2$ and $g_2(t) = t - 1/2$ for $1/2 \leq t \leq 1$. Let $\mathcal{F} = \{g_1, g_2\}$ and $f : [0, 1] \to 2^{[0,1]}$ be the set-theoretic union of g_1 and g_2. Then, $G(f^2) = Id \cup (1 - Id)$ and $\varprojlim f^2$ is not homeomorphic to $\varprojlim f$. (The graph of f is shown in Fig. 2.10; the graph of f^2 is shown in Fig. 2.6.)

Proof. In our discussion of a model for $M = \varprojlim f$ in Example 2.11, we observed that simple closed curves abound in this inverse limit. However, for completeness, we provide a specific simple closed curve lying in M. Let h be the sequence $g_2, g_1, g_2, g_1, \ldots$ and $A_1 = \varprojlim h$. Let $A_2 = \varprojlim g_1$. Let k be the sequence $g_1, g_2, g_1, g_2, \ldots$ and $A_3 = \varprojlim k$. Finally, let $A_4 = \varprojlim g_2$. It is not difficult to see that A_i is an arc for $1 \leq i \leq 4$. For instance, that A_1 is an arc follows from the fact that A_1 is homeomorphic to $\varprojlim g_2 \circ g_1$ which in turn is homeomorphic to $\varprojlim h$ where $h = Id_{[0,1/2]}$. Furthermore, $A_1 \cap A_2 = \{(1/2, 1, 1/2, 1, \ldots)\}$, $A_2 \cap A_3 = \{(1, 1/2, 1, 1/2, \ldots)\}$, $A_3 \cap A_4 = \{1/2, 0, 1/2, 0, \ldots)\}$, $A_4 \cap A_1 = \{(0, 1/2, 0, 1/2, \ldots)\}$, $A_2 \cap A_4 = \emptyset$, and $A_1 \cap A_3 = \emptyset$. Thus, $A_1 \cup A_2 \cup A_3 \cup A_4$ is a simple closed curve lying in M.

A calculation of f^2 shows that this is the bonding function in Example 2.7, i.e., $G(f^2)$ is the union of the two maps Id and $1 - Id$. That inverse limit is a the cone over the Cantor set, a continuum containing no simple closed curve. □

It is not difficult to see that, for the bonding function f from Example 3.2, $\varprojlim f^2$ is homeomorphic to a plane continuum, while we observed in Example 2.11 that $\varprojlim f$ cannot be embedded in the plane. This provides an alternate proof that $\varprojlim f$ and $\varprojlim f^2$ cannot be homeomorphic.

We close this section with a simple observation related to Problem 6.10. If $f : [0, 1] \to 2^{[0,1]}$ is an upper semicontinuous function, f is said to be *idempotent* provided $f = f^2$. The function f that is the union of Id and $1 - Id$ is idempotent, while the function from Example 3.2 is not idempotent. It is not difficult to show that if $f : [0, 1] \to 2^{[0,1]}$ is idempotent, then $f^n = f$ for each positive integer n. Thus, we have the following theorem.

Theorem 3.2. *If* $f : [0, 1] \to 2^{[0,1]}$ *is upper semicontinuous and* f *is idempotent, then* $\varprojlim f$ *is homeomorphic to* $\varprojlim f^n$ *for each* $n \in \mathbb{N}$.

To end this section, we mention that the function in Example 2.27 is idempotent, its graph is connected, and its inverse limit is not connected.

3.3   Bonding Functions Vis-À-Vis Projections

In inverse limits with mappings, the bonding functions and the projections interact in the following way.

Theorem 3.3. *Suppose* f *is a sequence of mappings of* $[0, 1]$ *into* $[0, 1]$ *and* $M = \varprojlim f$. *If* $H \subseteq M$, *then* $\pi_i(H) = f_i(\pi_{i+1}(H))$ *for each positive integer* i.

This fails for inverse limits with set-valued functions as may be seen by the following example, although it is true that $\pi_i(H) \subseteq f_i(\pi_{i+1}(H))$ for each positive integer i (see Theorem 3.4 below).

Example 3.3. Let $f : [0, 1] \to 2^{[0,1]}$ be given by $f(t) = [0, 1]$ for each $t \in [0, 1]$. Then, for $H = [0, 1/2]^\infty$, we have $H \subseteq \varprojlim f$ and, if $i \in \mathbb{N}$, $\pi_i(H) = [0, 1/2]$, but $f_i(\pi_{i+1}(H)) = [0, 1]$.

The following theorem follows directly from Theorem 1.10.

Theorem 3.4. *Suppose* f *is a sequence of upper semicontinuous functions from* $[0, 1]$ *into* $2^{[0,1]}$ *and* $M = \varprojlim f$. *If* $m, n \in \mathbb{N}$ *with* $m \leq n$ *and* $H \subseteq M$, *then* $\pi_m(H) \subseteq f_{mn}(\pi_n(H))$.

In case f is a sequence of upper semicontinuous functions if $H \subseteq \varprojlim f$ and i is a positive integer such that $f_i|\pi_{i+1}(H)$ is a mapping, it is true that $\pi_i(H) = f_i(\pi_{i+1}(H))$.

3.4   The Closed Subset Theorem

Another quite valuable tool in the theory of inverse limits with mappings is the closed subset theorem. For sequences of mappings on $[0, 1]$, its statement is found below. In the statement of Theorem 3.5 as well as in Example 3.4, we employ the following notation. If H is a closed subset of Q, we denote $\pi_i(H)$ by H_i. A proof of Theorem 3.5 can be found in [4, Theorem 159, p. 116].

Theorem 3.5. *If* f *is a sequence of mappings of* $[0, 1]$ *into* $[0, 1]$ *and* H *is a closed subset of* $\varprojlim f$, *then* $H = \varprojlim g$ *where* $g_i = f_i|H_{i+1}$ *for each positive integer* i.

Numerous examples show that this theorem fails to hold in general for inverse limits on $[0, 1]$ with set-valued functions. One such example is Example 2.13 which we revisit in our next example.

Example 3.4. Let $f : [0, 1] \to C([0, 1])$ be given by $f(t) = \{0, t\}$ for $0 \le t \le 1$. Then, $H = \{x \in \underleftarrow{\lim} f \mid x_i = x_1 \text{ for each positive integer } i\}$ is a closed proper subset of $\underleftarrow{\lim} f$ such that $H_n = [0, 1]$ for each positive integer n.

Our next section is devoted to addressing the failure of Theorem 3.5 for inverse limits with set-valued functions in the case that closed subset of the inverse limit is a continuum that projects onto each of infinitely many factor spaces. A deeper study of compact subsets of the inverse limit that are the inverse limit of their projections was conducted by Alexander Cornelius in his dissertation at Baylor University [1] as well as in a subsequent study of Brian Williams in his Baylor dissertation [6].

3.5   The Full Projection Property

One of the principal uses of Theorem 3.5 in inverse limits with mappings is to conclude that a subcontinuum of an inverse limit that projects onto each factor space is the entire inverse limit. Although examples, including Example 3.4, exist showing that this does not always hold for set-valued functions, for some inverse limits with set-valued functions, subcontinua that project onto the full factor space for infinitely many integers n must be the entire inverse limit. Those for which this is true are said to satisfy the full projection property. Specifically, suppose f is a sequence of upper semicontinuous functions from $[0, 1]$ into $2^{[0,1]}$ and $M = \underleftarrow{\lim} f$. Then, M has the *full projection property* provided it is true that if H is a subcontinuum of M such that $\pi_n(H) = [0, 1]$ for infinitely many positive integers n then $H = M$.

Our next example has the full projection property. This was first shown by Scott Varagona [5]. Our proof is based on the same principles as his, but it is different. It may be worth noting that the proof does not make use of the connectedness of H.

Example 3.5 (Varagona). Let $f : [0, 1] \to C([0, 1])$ be the function whose graph is the union of straight line intervals joining $(1/2^n, 0)$ and $(1/2^{n-1}, 1)$ for all odd positive integers, straight line intervals joining $(1/2^{n-1}, 0)$ and $(1/2^n, 1)$ for all even positive integers, and the straight line interval joining $(0, 0)$ and $(0, 1)$ (a graph homeomorphic to a $\sin(1/x)$-curve; see Fig. 3.2). Then, $M = \underleftarrow{\lim} f$ has the full projection property.

Proof. Suppose H is a subcontinuum of M such that $\pi_i(H) = [0, 1]$ for infinitely many integers i. One key to proving that $H = M$ is an observation made by Varagona in his proof.

(1) *If $p \in M$ and $p_i > 0$ for each positive integer i, then $p \in H$.*

To see this, let n be a positive integer. There is an integer $m > n$ such that $\pi_m(H) = [0, 1]$. Therefore, there is a point $x \in M$ such that $x_m = p_m$. Because $p_m > 0$, $f(p_m) = p_{m-1}$, so $x_{m-1} = p_{m-1} > 0$. Continuing inductively, we see that $x_i = p_i$ for $1 \le i \le m$. Thus, $d(x, p) < 1/2^m \le 1/2^n$. It follows that p is a limit point of H. Because H is closed, $p \in H$.

Fig. 3.2 The graph of the
bonding function in
Example 3.5

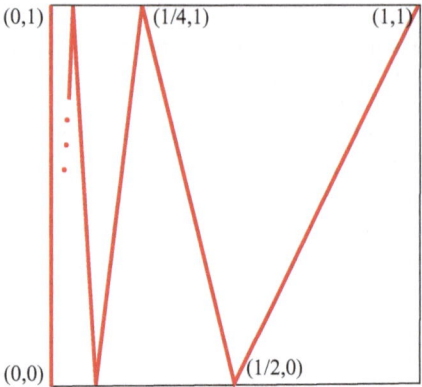

The remainder of the proof is devoted to showing that if $x \in M$ and $\varepsilon > 0$, there is a point $p \in H$ such that $d(p, x) < \varepsilon$. Toward this end and employing the notation that f^0 denotes the identity, we first observe the following.

(2) *Suppose $z > 0$, $w \in [0, 1]$, and k is a positive integer such that $f^k(z) = 0$, but $f^i(z) > 0$ for $1 \leq i < k$. If $\varepsilon > 0$, there is a number $t > 0$ such that $|f^i(t) - f^i(z)| < \varepsilon$, $f^i(t) > 0$ for $0 \leq i \leq k$, and $f^{k+1}(t) = w$.*

To see this, note that f^i is continuous at z for $1 \leq i \leq k$. So, if $1 \leq i \leq k$, there is a positive number δ_i such that if $|t - z| < \delta_i$, then $|f^i(t) - f^i(z)| < \varepsilon$. The fact that $f^i(z) > 0$ along with the continuity of f^i at z yields that for $1 \leq i < k$, there is a positive number δ_i' such that if $|t - z| < \delta_i'$ then $f^i(t) > 0$. Furthermore, because $f^{k-1}(z) > 0$, z is an isolated zero for f^k. Thus, because f^k is continuous at z, there is a positive number δ_k' such that if $0 < |t - z| < \delta_k'$, then $f^k(t) > 0$. Let $\delta = \min\{\delta_1, \delta_2, \ldots, \delta_k, \delta_1', \delta_2', \ldots, \delta_k', \varepsilon\}$ and let J be the open interval $(z - \delta, z + \delta)$. Because $f^k(J)$ is a connected set containing 0, there is a positive integer m such that $[1/2^m, 1/2^{m-1}] \subseteq f^k(J)$. Thus there is a point $s \in [1/2^m, 1/2^{m-1}]$ such that $f(s) = w$. There is a point $t \in J$ such that $f^k(t) = s$. Then $f^{k+1}(t) = w$. Because $s > 0$, $t \neq z$, thus $|f^i(t) - f^i(z)| < \varepsilon$ and $f^i(t) > 0$ for $1 \leq i \leq k$.

Before we undertake the last part of the proof, we include one additional item that is used in the proof.

(3) *If $j \in \mathbb{N}$ and $x_1, x_2, \ldots, x_j$ are points of $[0, 1]$ such that $f(x_{i+1}) = x_i$ for $1 \leq i < j$, then there is a point y of M such that $y_i = x_i$ for $1 \leq i \leq j$ and $y_i > 0$ for $i > j$.*

Construct $y \in M$ as follows. Let $y_i = x_i$ for $1 \leq i \leq j$. There is a point $t > 0$ such that $f(t) = x_j$. Let y_{j+1} be such a point t. There is a point $s > 0$ such that $f(s) = y_{j+1}$. Let y_{j+2} be such a point s. Continuing in this manner, we inductively construct y.

To complete the proof we show the following which, in light of (1), shows that every point of M is a limit point of the closed set H and, hence, is in H.

(4) *If $x \in M$ and $\varepsilon > 0$, there is a point $p \in M$ such that every coordinate of p is positive and $d(p, x) < \varepsilon$.*

To prove this, suppose $\varepsilon > 0$ and $x \in M$. There is a positive integer n such that $\sum_{i \geq n} 1/2^i < \varepsilon/4$. We proceed by induction on the number of times 0 appears among the first n coordinates of x. If that number is zero, let p be the point of M produced by (3) using $j = n$ and the points $x_1, x_2, \ldots, x_n$. Then, every coordinate of p is positive. Because $p_i = x_i$ for $1 \leq i \leq n$, $d(p, x) < \varepsilon/4$.

Inductively, suppose $k \geq 0$ is an integer such that if y is a point of M having k of its first n coordinates 0, then there is a point of M having all positive coordinates and distance less than $\varepsilon/2$ from y. Let x be a point of M having $k + 1$ of its first n coordinates 0. We consider two cases: (a) $x_n = 0$ and (b) $x_n > 0$. If $x_n = 0$, there is a positive integer m such that $[1/2^m, 1/2^{m-1}] \subseteq [0, \varepsilon/2)$. Thus, there is a point $t \in [1/2^m, 1/2^{m-1}]$ such that $f(t) = x_{n-1}$. Using the points $x_1, x_2, \ldots, x_{n-1}, t$ in (3), there is a point z of M such that $z_i = x_i$ for $1 \leq i \leq n-1$, $z_n = t$, and $z_i > 0$ for $i > n$. Observe that $d(z, x) < \varepsilon/4 + \varepsilon/4 < \varepsilon/2$. Because z has only k coordinates that are 0 among its first n coordinates, by the inductive hypothesis, there is a point p of M having all coordinates positive such that $d(p, z) < \varepsilon/2$. Thus, $d(p, x) < \varepsilon$. Suppose (b) holds. Then there is a positive integer $j < n$ such that $x_j = 0$, but $x_i > 0$ for $j < i \leq n$. Using $\varepsilon/4$ in (2) with $z = x_n$ and $k = n - j$, we obtain a number $t > 0$ such that $f^i(t) > 0$ and $|f^i(t) - f^i(x_n)| < \varepsilon/4$ for $0 \leq i \leq n - j$, whereas $f^{n-j+1}(t) = x_{j-1}$. Using the points $x_1, x_2, \ldots, x_{j-1}, f^{n-j}(t), \ldots, t$ in (3), we obtain a point $z \in M$ such that z has k coordinates 0 among its first n coordinates. Because $f^l(x_n) = x_{n-l}$ for $0 \leq l \leq j$ and, thus, $|f^i(t) - x_{n-i}| < \varepsilon/4$ for $0 \leq i \leq n - j$, it is not difficult to see that $d(z, x) < \varepsilon/4$. By the inductive hypothesis, there is a point p of M having all coordinates positive such that $d(p, z) < \varepsilon/2$. Thus, $d(p, x) < \varepsilon$.   ⊔

A simpler function that also has an inverse limit having the full projection property is the function in our next example. This example is also found in the literature, [2, Example 3.4, p. 365]. We revisit Examples 3.5 and 3.6 in Sect. 3.6 where both inverse limits are shown to be indecomposable.

As with Example 3.5, we provide a proof that the inverse limit in Example 3.6 has the full projection property that differs from the one in the literature. Recall that if f is a sequence of set-valued functions such that $f_i : [0, 1] \to 2^{[0,1]}$ for each $i \in \mathbb{N}$, then $G'(f_1, f_2, \ldots, f_n) = \{x \in [0, 1]^{n+1} \mid x_i \in f_i(x_{i+1}) \text{ for } 1 \leq i \leq n\}$. For convenience of notation, we let $G'_n = G(f_1, f_2, \ldots, f_n)$. We begin with a lemma. We leave its proof to the reader.

Lemma 3.1. *Suppose f is a sequence of set-valued functions such that $f_i : [0, 1] \to 2^{[0,1]}$ for each positive integer i. If $n \in \mathbb{N}$, then $G'_{n+1} = \{x \in [0, 1]^{n+2} \mid (x_1, x_2, \ldots, x_{n+1}) \in G'_n \text{ and } x_{n+2} \in f_{n+1}^{-1}(x_{n+1})\}$.*

Example 3.6. Let $f : [0, 1] \to 2^{[0,1]}$ be given by $f(t) = 2t$ for $0 \leq t < 1/2$, $f(1/2) = [0, 1]$, and $f(t) = 2t - 1$ for $1/2 < t \leq 1$. Then, $\varprojlim f$ has the full projection property (see Fig. 3.3 for the graph of f).

Fig. 3.3 The graph of the
bonding function in
Example 3.6

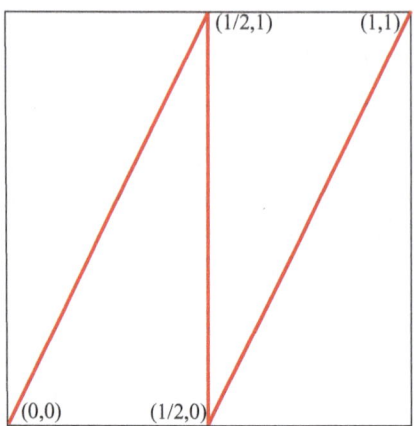

Proof. We first show that if $n \in \mathbb{N}$, then G'_n is an arc in $[0, 1]^{n+1}$ from $(0, 0, \ldots, 0)$
to $(1, 1, \ldots, 1)$. We proceed by induction.

Because $G'_1 = G(f)^{-1}$ and $G(f)$ is an arc, G'_1 is an arc having endpoints $(0, 0)$
and $(1, 1)$. Suppose $k \in \mathbb{N}$ such that G'_k is an arc from $(0, 0, \ldots, 0)$ to $(1, 1, \ldots, 1)$.
By Lemma 3.1, $G'_{k+1} = \{x \in [0, 1]^{k+2} \mid (x_1, x_2, \ldots, x_{k+1}) \in G'_k$ and $x_{k+2} \in$
$f^{-1}(x_{k+1})\}$. Observe that f^{-1} is the union of three mappings f_1, f_2, and f_3 given
by $f_1(t) = t/2$, $f_2(t) = 1/2$ and $f_3(t) = (t + 1)/2$ for $0 \leq t \leq 1$. Thus, if
$h_i : G'_k \to G'_{k+1}$ is the homeomorphism given by $h_i(x) = (x_1, x_2, \ldots, x_k, f_i(x_k))$
for $i = 1, 2, 3$, $G'_{k+1} = h_1(G'_k) \cup h_2(G'_k) \cup h_3(G'_k)$. Note that $\alpha_1 = h_1(G'_k)$ is an arc
in $[0, 1]^{n+2}$ with endpoints $(0, 0, \ldots, 0)$ and $(1, 1, \ldots, 1, 1/2)$, $\alpha_2 = h_2(G'_k)$ is an
arc in $[0, 1]^{n+1}$ with endpoints $(0, 0, \ldots, 0, 1/2)$ and $(1, 1, \ldots, 1, 1/2)$, and $\alpha_3 =$
$h_3(G'_k)$ is an arc in $[0, 1]^{n+1}$ with endpoints $(0, 0, \ldots, 0, 1/2)$ and $(1, 1, \ldots, 1)$.
Furthermore, $\alpha_1 \cap \alpha_2 = \{(1, 1, \ldots, 1, 1/2)\}$, $\alpha_2 \cap \alpha_3 = \{(0, 0, \ldots, 0, 1/2)\}$, and
$\alpha_1 \cap \alpha_3 = \emptyset$. It follows that G'_{k+1} is an arc with endpoints $(0, 0, \ldots, 0)$ and
$(1, 1, \ldots, 1)$.

Suppose H is a subcontinuum of $M = \varprojlim f$ such that $\pi_i(H) = [0, 1]$ for
infinitely many positive integers i. Suppose p is a point of M and $n \in \mathbb{N}$. We show
that there is a point q of H such that $q_j = p_j$ for $1 \leq j \leq n$. To see this we
consider three cases: $p = (0, 0, 0, \ldots)$, $p = (1, 1, 1, \ldots)$, and $p \in M - \{(0, 0, 0,$
$\ldots), (1, 1, 1, \ldots)\}$. Suppose $p = (0, 0, 0, \ldots)$. There is an integer $m \geq n$ such that
$\pi_m(H) = [0, 1]$, so there is a point $q \in H$ such that $q_m = 0$. Because $f(0) = 0$,
$q_j = 0$ for $1 \leq j \leq m$. The case that $p = (1, 1, 1, \ldots)$ is similar. Suppose p
is a point of $M - \{(0, 0, 0, \ldots), (1, 1, 1, \ldots)\}$. There is a positive integer $m \geq n$
such that $p_{m+1} \notin \{0, 1\}$ and $\pi_{m+1}(H) = [0, 1]$. Then, $(p_1, \ldots, p_{m+1})$ is in the arc
G'_m and $G'_m - \{(p_1, \ldots, p_{m+1})\} = A_{m,0} \cup A_{m,1}$, where $A_{m,0}$ and $A_{m,1}$ are mutually
separated with $(0, 0, \ldots, 0) \in A_{m,0}$ and $(1, 1, \ldots, 1) \in A_{m,1}$. Thus, H intersects
the two mutually separated sets $A_{m,0} \times [0, 1]^\infty$ and $A_{m,1} \times [0, 1]^\infty$ so H, contains a
point in the boundary of each of them. Consequently, H contains a point q such that

$q_j = p_j$ for $1 \le j \le m$. In each case, there is a point $q \in H$ such that $q_j = p_j$ for $1 \le j \le n$. Thus, there is a point q of H such that $d(q, p) < 2^{-n}$. It follows that $p \in \overline{H}$, so $H = \varprojlim f$.   □

A characterization of the full projection property is included in Brian Williams' dissertation [6], although it is not in terms of properties of the bonding functions. One research problem related to this section is Problem 6.18. The reader will notice that our proofs that Examples 3.5 and 3.6 have the full projection property are quite different. This leads to Problem 6.26 which may be more tractable than Problem 6.18.

3.6   Indecomposability and the Two-Pass Condition

A continuum is *decomposable* provided it is the union of two proper subcontinua and is *indecomposable* otherwise. Although many examples of indecomposable continua are known to continuum theorists, it is not obvious that there are nondegenerate indecomposable continua. However, we shall soon see that such continua exist.

In the presence of the full projection property the two-pass condition (defined below) is a sufficient condition for indecomposability of an inverse limit when the factor spaces belong to the collection of all arcs and simple n-ods. We show this for the interval $[0, 1]$ in Theorem 3.6. Without the full projection property, an inverse limit with a single bonding function may be decomposable even if the bonding function satisfies the two-pass condition, e.g., for $f : [0, 1] \rightarrow 2^{[0,1]}$ given by $f(t) = [0, 1]$ for $0 \le t \le 1$ as in Example 1.1 where the inverse limit is the Hilbert cube, a decomposable continuum.

If $f : [0, 1] \rightarrow 2^{[0,1]}$, f is said to satisfy the *two-pass condition* provided there exist two mutually exclusive connected open subsets U and V of $[0, 1]$ such that $f|U$ and $f|V$ are mappings and $\overline{f(U)} = \overline{f(V)} = [0, 1]$. The following theorem appears in [2, Theorem 4.3, p. 370] in a more general form.

Theorem 3.6. *If f is a sequence of upper semicontinuous functions from $[0, 1]$ into $2^{[0,1]}$, f_i satisfies the two-pass condition for each positive integer i, and $\varprojlim f$ has the full projection property, then $\varprojlim f$ is indecomposable.*

Proof. Let $M = \varprojlim f$ and suppose M is the union of two proper subcontinua H and K. Because M has the full projection property, there is a positive integer n such that if $j \ge n$, then $\pi_j(H) \ne [0, 1]$ and $\pi_j(K) \ne [0, 1]$. There exist two mutually exclusive connected open subsets U and V of $[0, 1]$ such that $f_n|U$ and $f_n|V$ are mappings and $\overline{f_n(U)} = \overline{f_n(V)} = [0, 1]$. However, because $\pi_{n+1}(H)$ and $\pi_{n+1}(K)$ are intervals whose union is $[0, 1]$ and U and V are mutually exclusive open intervals, one of U and V is a subset of one of $\pi_{n+1}(H)$ and $\pi_{n+1}(K)$. Suppose $U \subseteq \pi_{n+1}(H)$. Because $f_n|U$ is a mapping, if $t \in f_n(U)$, then there is a point

$s \in U$ such that $t = f_n(s)$. But, $s \in \pi_{n+1}(H)$ so there is a point $p \in H$ such that $p_{n+1} = s$. Then, $p_n = t$ so $t \in \pi_n(H)$, so $f_n(U) \subseteq \pi_n(H)$. Consequently, $\pi_n(H) = [0, 1]$. □

In the previous proof, we went to some lengths to show that, if $U \subseteq \pi_{n+1}(H)$, then $f_n(U) \subseteq \pi_n(H)$ by using the hypothesis that $f_n|U$ is a mapping. We saw in Example 3.3 that this does not always hold for set-valued functions even if $U = \pi_{n+1}(H)$.

Example 3.7 (An indecomposable continuum). The inverse limit M in Example 3.5 is an indecomposable continuum. (The graph of the bonding function is homeomorphic to a $\sin(1/x)$-curve; see Fig. 3.2.)

Proof. It follows from Theorem 2.7 that M is a continuum. The function f is easily seen to satisfy the two-pass condition. We showed in Example 3.5 that M has the full projection property, so it follows from Theorem 3.6 that M is indecomposable. □

Of course, the indecomposability of the inverse limit in Example 3.7 depends on the embedding of the $\sin(1/x)$-curve in $[0, 1]^2$.

Example 3.8. If f is the function from Example 3.5 and $g : [0, 1] \to 2^{[0,1]}$ is given by $g(t) = t$ for $0 \le t \le 1/2$ and $g(t) = (2 - f(2 - 2t))/2$ for $1/2 < t \le 1$, then $G(g)$ is homeomorphic to a $\sin(1/x)$-curve but $\varprojlim g$ is decomposable.

In a similar manner to Example 3.7, but in this case relying on Example 3.6, we obtain the following example.

Example 3.9 (An indecomposable continuum). The inverse limit M in Example 3.6 is an indecomposable continuum. (The bonding function $f : [0, 1] \to 2^{[0,1]}$ is given by $f(t) = 2t$ for $0 \le t < 1/2$, $f(1/2) = [0, 1]$, and $f(t) = 2t - 1$ for $1/2 < t \le 1$; see Fig. 3.3.)

Finally, we provide an example due to Varagona of a function satisfying the two-pass condition, but its inverse limit does not have the full projection property. Varagona's example is constructed by tacking a vertical line at 0 onto the full tent map as seen in our next example.

Example 3.10 (A function satisfying the two-pass condition having an inverse limit without the full projection property). Let $f : [0, 1] \to 2^{[0,1]}$ be the function given by $f(0) = [0, 1]$, $f(t) = 2t$ for $0 < t < 1/2$, and $f(t) = 2 - 2t$ for $1/2 < t \le 1$. Then f satisfies the two-pass condition but $\varprojlim f$ does not have the full projection property. (See Fig. 3.4 for the graph of f.)

Proof. Let $M = \varprojlim f$. By choosing $U = (0, 1/2)$ and $V = (1/2, 1)$, we see that f satisfies the two-pass condition. If $g : [0, 1] \to [0, 1]$ is the map given by $g(t) = 2t$ for $0 \le t \le 1/2$ and $g(t) = 2 - 2t$ for $1/2 < t \le 1$, we see $H = \varprojlim g$ is a proper subcontinuum of M that projects onto $[0, 1]$ for each positive integer n. □

Fig. 3.4 The graph of the bonding function in Example 3.10

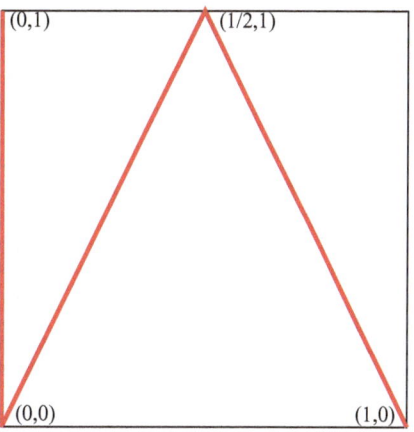

3.7 Bennett's Theorem

In [2, Example 3.5], we introduced the following example to show that a graph could be irreducible from $\{0\} \times [0, 1]$ to $\{1\} \times [0, 1]$ while failing to have the full projection property. The example is somewhat interesting for other reasons because the inverse limit is not the closure of a topological ray (i.e., the image of the nonnegative real numbers under a homeomorphism) with remainder (the complement of the ray in its closure) homeomorphic to the inverse limit on $[1/2, 1]$. In its simplest form, Bennett's Theorem yields that if $g : [1/2, 1] \to [1/2, 1]$ is a mapping with $g(1/2) - 1$ and $f : [0, 1] \to [0, 1]$ is a mapping such that $f|[1/2, 1] = g$, $f(t) = 2t$ for $0 \le t \le 1/2$, then $\varprojlim f$ is the closure of a ray with remainder homeomorphic to $\varprojlim g$. See [4, Theorem 19, p. 12] for a stronger version of this theorem. Example 3.11 thus shows that, in general, we cannot expect a version of Bennett's Theorem to hold for inverse limits with set-valued functions.

Example 3.11 (Bennett's Theorem fails). Let $f : [0, 1] \to 2^{[0,1]}$ be given by $f(t) = 2t$ for $0 \le t < 1/2$, $f(1/2) = [1/2, 1]$, and $f(t) = t$ for $1/2 < t \le 1$. Then the inverse limit is not the closure of a ray with remainder $\varprojlim g$ where $g = f|[1/2, 1]$ (Fig. 3.5).

Proof. Let $M = \varprojlim f$. It follows from Theorem 2.7 that M is a continuum, but we wish to construct a model for the continuum from which one can see that M is not the closure of a ray with remainder $\varprojlim g$ where $g = f|[1/2, 1]$. Let $A_0 = \{x \in M \mid x_1 \in [1/2, 1]$ and $x_j = x_1$ for each positive integer $j\}$. For each positive integer n, let $A_n = \{x \in M \mid x_1 \in [1/2, 1], x_j = x_1$ for $1 \le j \le n$, and $x_j = 1/2$ for $j > n\}$. Then $F = \bigcup_{i>0} A_i$ is a fan lying in M having vertex $p = (1/2, 1/2, 1/2, \ldots)$. (This fan is $\varprojlim g$ and is homeomorphic to the inverse limit in Example 2.14.)

Let $A(0, 0) = \{x \in M \mid x_1 \in [0, 1/2]$ and $x_{j+1} = x_j/2$ for $j \ge 1\}$. Suppose n is a positive integer. Let $A(n, 0) = \{x \in M \mid x_1 \in [1/2, 1], x_j = x_1$ for $1 \le j \le n$,

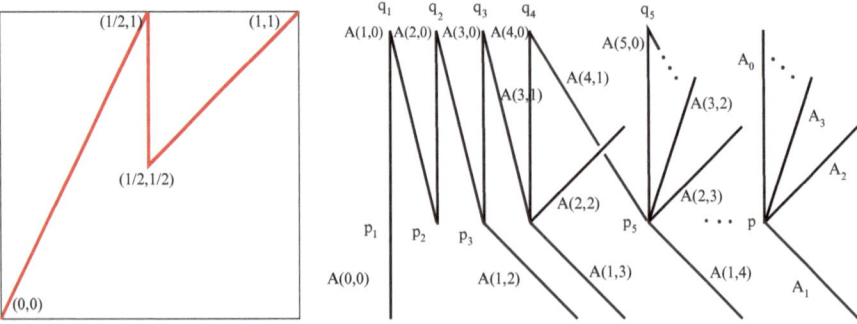

Fig. 3.5 The graph of the bonding function and a model of the inverse limit in Example 3.11

and $x_{j+1} = x_j/2$ for $j \geq n\}$. If m is a positive integer, let $A(n,m) = \{x \in M \mid x_1 \in [1/2, 1], x_j = x_1$ for $1 \leq j \leq n, x_j = 1/2$ for $n < j \leq n+m$, and $x_{j+1} = x_j/2$ for $j \geq n+m\}$. Note that if $x \in M - F$, then there exist an integer $n \geq 0$ and an integer $m \geq 0$ such that $x \in A(n,m)$.

For $n \geq 3$, let $T_n = A(n,0) \cup A(n-1,1) \cup A(n-2,2) \cup \cdots \cup A(1, n-1)$. Then, T_n is an n-od lying in M having vertex p_n where p_n is the point of M such that $\pi_j(p_n) = 1/2$ for $1 \leq j \leq n$ and $\pi_{j+1}(p_n) = \pi_j(p_n)/2$ for $j \geq n$. Let q_n be the point of M such that $\pi_j(q_n) = 1$ for $1 \leq j \leq n$ and $\pi_{j+1}(q_n) = \pi_j(q_n)/2$ for $j \geq n$. Observe that, for $n \geq 3$, $T_n \cap T_{n+1} = \{q_n\}$ because $q_n \in A(n,0) \cap A(n,1)$ and $A(n,0) \subseteq T_n$, while $A(n,1) \subseteq T_{n+1}$.

Because $A(0,0) \cap A(1,0) = \{p_1\}$, $A(1,0) \cap A(1,1) = \{q_1\}$, $A(1,1) \cap A(2,0) = \{p_2\}$, and $A(2,0) \cap A(2,1) = \{q_2\}$, we see that $S = A(0,0) \cup A(1,0) \cup A(1,1) \cup A(2,0) \cup (\bigcup_{i \geq 3} T_i)$ is connected. Further, $\overline{S} - S = F$ and $M = \overline{S}$. However, no topological ray lying in M can have M as its closure. A maximal ray having a closure that is not locally connected is $A(0,0) \cup A(1,0) \cup A(1,1) \cup A(2,0) \cup A(2,1) \cup A(3,0) \cup \cdots$. $\square$

References

1. Cornelius, A.N.: Inverse limits of set-valued functions. Ph.D. dissertation, Baylor University (2008)
2. Ingram, W.T.: Inverse limits with upper semi-continuous bonding functions: Problems and partial solutions. Topology Proc. **36**, 353–373 (2010)
3. Ingram, W.T., Mahavier, W.S.: Inverse limits of upper semi-continuous set valued functions. Houston J. Math. **32**, 119–130 (2006)
4. Ingram, W.T., Mahavier, W.S.: Inverse limits: From continua to Chaos. In: Developments in Mathematics, vol. 25. Springer, New York (2012)
5. Varagona, S.: Inverse limits with upper semi-continuous bonding functions and indecomposability. Houston J. Math. **37**, 1017–1034 (2011)
6. Williams, B.R.: Indecomposability in inverse limits. Ph.D. dissertation, Baylor University (2010)

Chapter 4
Mapping Theorems

Abstract In this chapter, we include some theorems on mappings of inverse limit spaces. Although the subsequence theorem for inverse limits with mappings does not hold in general for inverse limits with set-valued functions, there is a version for upper semicontinuous functions that gives a mapping between inverse limits including, specifically, a mapping of $\varprojlim f$ onto $\varprojlim f^2$ for inverse limits with a single bonding function. The shift homeomorphisms between inverse limits with mappings also do not carry over as homeomorphisms to the set-valued case. Instead, one shift is a mapping and the other is a set-valued function. A generalized conjugacy theorem rounds out this chapter.

4.1 Introduction

In Chap. 2 we discussed the effect on the inverse limit of topological conjugacy between the bonding functions in an inverse limit sequence. This is a form of a mapping theorem in that the conjugacy produces a homeomorphism between inverse limits. In this chapter we explore other mapping theorems between inverse limit sequences. Most of our discussion centers around theorems that hold because the projections from product spaces to their factor spaces (or products of their factor spaces) are continuous. Theorems 4.1 and 4.5 are from [2] but may also be found in [3].

4.2 A Subsequence Mapping Theorem

In this section we present a theorem on mappings induced by projections. Theorem 4.1 is the set-valued counterpart to the subsequence theorem for inverse limits with mappings. Unlike the situation with bonding mappings, in the case that the bonding functions are set-valued, the induced mapping is not a homeomorphism.

W.T. Ingram, *An Introduction to Inverse Limits with Set-valued Functions*, 59
SpringerBriefs in Mathematics, DOI 10.1007/978-1-4614-4487-9_4,
© W.T. Ingram 2012

Recall our notation that if f is an inverse limit sequence of set-valued functions and $i, j \in \mathbb{N}$ with $i < j$, then $f_{ij} = f_i \circ f_{i+1} \circ \cdots \circ f_{j-1}$, a set-valued function from the jth factor space into the closed subsets of the ith factor space.

Theorem 4.1. *Suppose X is a sequence of closed subsets of $[0, 1]$, and, for each positive integer i, $f_i : X_{i+1} \to 2^{X_i}$ is a surjective upper semicontinuous function. Suppose further that $n_1, n_2, n_3, \ldots$ is an increasing sequence of positive integers and g is a sequence of upper semicontinuous functions such that $g_i : X_{n_{i+1}} \to 2^{X_{n_i}}$, where $g_i = f_{n_i n_{i+1}}$. Then $F : \varprojlim f \twoheadrightarrow \varprojlim g$ given by $F(x) = \pi_{\{n_1, n_2, n_3, \ldots\}}(x)$ is a surjective mapping.*

Proof. Let $M = \varprojlim f$ and $N = \varprojlim g$. The continuity of F is a consequence of its definition. Because $F(x) = (x_{n_1}, x_{n_2}, x_{n_3}, \ldots)$, it is not difficult to verify that $x_{n_i} \in g_i(x_{n_{i+1}})$, so $F : M \to N$.

To see that F is surjective, suppose $y \in N$. By Theorem 1.7, there is a point $x^1 \in M$ such that $\pi_{n_1}(x^1) = y_1$. By Theorem 1.8 there is a point $z \in M$ such that $\pi_{n_1}(z) = y_1$ and $\pi_{n_2}(z) = y_2$. Let x^2 be a point of M such that $\pi_j(x^2) = \pi_j(x^1)$ for $j \leq n_1$ and $\pi_j(x^2) = \pi_j(z)$ for $j > n_1$. Thus, $\pi_{n_i}(x^2) = y_i$ for $i \leq 2$. Again, by Theorem 1.8 there is a point $w \in M$ such that $\pi_{n_2}(w) = y_2$ and $\pi_{n_3}(w) = y_3$. Let x^3 be a point of M such that $\pi_j(x^3) = \pi_j(x^2)$ for $j \leq n_2$ and $\pi_j(x^3) = \pi_j(w)$ for $j > n_2$. Observe that $\pi_{n_i}(x^3) = y_i$ for $i \leq 3$. Proceeding inductively, we obtain a sequence $x^1, x^2, x^3, \ldots$ of points of M such that, if k is a positive integer, then $\pi_j(x^{k+1}) = \pi_j(x^k)$ for $j \leq n_k$, and $\pi_{n_i}(x^k) = y_i$ for $i \leq k$. The sequence $x^1, x^2, x^3, \ldots$ converges to a point $x \in M$ such that $\pi_{n_i}(x) = y_i$ for each positive integer i, i.e., $F(x) = y$. $\square$

In the case that the bonding functions are mappings in Theorem 4.1, it is known that the induced mapping F is 1–1. Even with a single set-valued bonding function, F need not be 1–1; for, otherwise, F would be a homeomorphism, but we saw in Example 3.2 that $\varprojlim f$ is not necessarily homeomorphic to $\varprojlim f^2$. Next, we revisit that example.

Example 4.1. Let $f : [0, 1] \to 2^{[0,1]}$ be the function given by $f(t) = \{1/2 + t, 1/2 - t\}$ for $0 \leq t \leq 1/2$, and $f(t) = \{3/2 - t, t - 1/2\}$ for $1/2 < t \leq 1$ (i.e., f is the function from Example 2.11). Then, $G(f^2) = Id \cup (1 - Id)$ (i.e., f^2 is the function from Example 2.7). There is a mapping from the Hurewicz continuum, $\varprojlim f$, onto the Cantor fan, $\varprojlim f^2$ (see Fig. 4.1 for graphs of f and f^2).

Proof. Let $n_1, n_2, n_3, \ldots$ be the sequence of odd integers and use Theorem 4.1. $\square$

Another example making use of Theorem 4.1 is Example 4.2 below. As in the previous example, there is a map from $\varprojlim f$ onto $\varprojlim f^2$. In this example the induced map is actually a homeomorphism, and $\varprojlim f$ is homeomorphic to $\varprojlim f^n$ for each $n \in \mathbb{N}$. (See Problem 6.10.)

Example 4.2. Let $f : [0, 1] \to C([0, 1])$ be given by $f(t) = t$ for $0 \leq t < 1/2$, $f(1/2) = [1/2, 1]$, and $f(t) = 1 - t$ for $1/2 < t \leq 1$ (f is the function from Example 2.21). Then $f^2 : [0, 1] \to C([0, 1])$ is given by $f^2(t) = t$ for $0 \leq t < 1/2$,

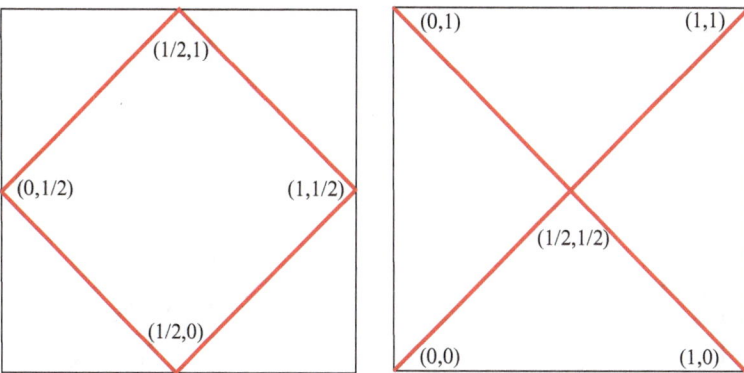

Fig. 4.1 Graphs of the functions f and f^2 in Example 4.1

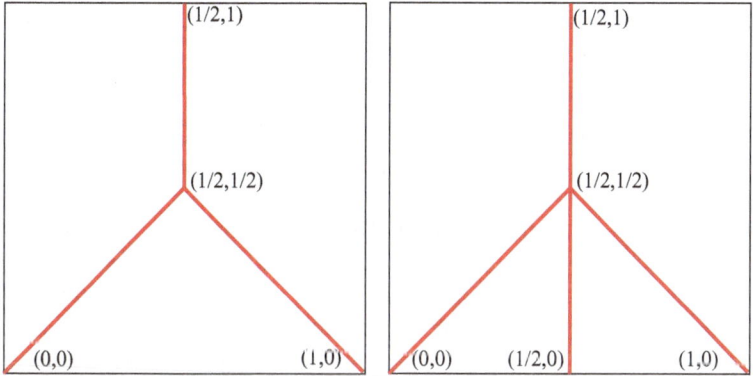

Fig. 4.2 Graphs of the functions f and f^2 in Example 4.2

$f^2(1/2) = [0, 1]$, and $f^2(t) = 1 - t$ for $1/2 < t \leq 1$; $\varprojlim f$ and $\varprojlim f^2$ are both fans, and there is a mapping from $\varprojlim f$ onto $\varprojlim f^2$ (see Fig. 4.2 for the graphs of f and f^2 and Fig. 4.3 for models of the inverse limits with $\varprojlim f$ pictured on the left). In fact, $\varprojlim f$ and $\varprojlim f^2$ are homeomorphic under the map from Theorem 4.1 induced by the sequence of odd positive integers. In addition, $f^3 = f^2$, and as a consequence, $\varprojlim f$ is homeomorphic to $\varprojlim f^n$ for each $n \in \mathbb{N}$.

Proof. Let $M = \varprojlim f$ and, as in the notation from Example 2.21, let $B_0 = \{x \in M \mid x_1 \in [0, 1/2] \text{ and } x_j = x_1 \text{ for } j > 1\}$. For $n \in \mathbb{N}$, let $B_n = \{x \in M \mid x_n \in [1/2, 1]\}$. Note that if $x \in B_n$, then $x_j = 1/2$ for $j > n$, and if $n > 1$, then $x_j \in [0, 1/2]$ for $j < n$. Further, if $n > 2$ and $x \in B_n$, then $x_n = 1 - x_{n-1}$ and $x_j = x_1$ for $j \leq n - 1$.

Let $N = \varprojlim f^2$. Because $G(f) \subseteq G(f^2)$, $M \subseteq N$, so the fan described in the previous paragraph is a subset of N. There are additional arcs lying in N that have

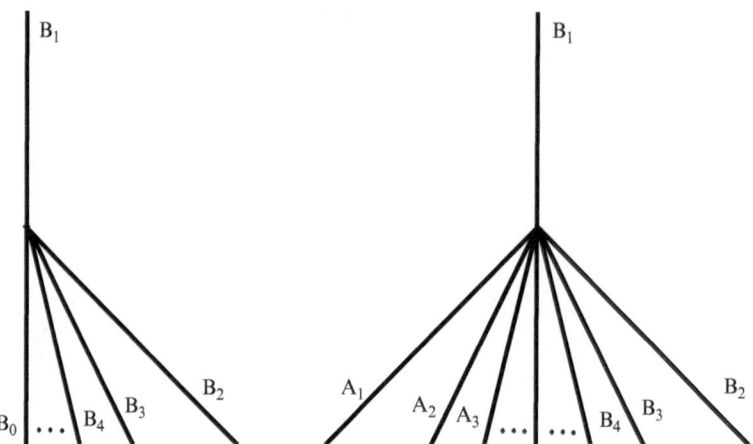

Fig. 4.3 Models of the inverse limits in Example 4.2 with the inverse limit of f^2 shown on the *right*

not yet been identified. For $n \in \mathbb{N}$, let $A_n = \{x \in N \mid x_1 \in [0, 1/2]$ and $x_j = x_1$ for $1 \le j \le n$ and $x_j = 1/2$ for $j > n\}$. If $x \in N - M$, there is an integer n such that $x \in A_n$.

Let $F : M \twoheadrightarrow N$ be the map from Theorem 4.1 induced by the sequence of odd positive integers. Note that $F|(B_0 \cup B_1)$ is the identity, and for $n \in \mathbb{N}$, $F|B_{2n}$ is a homeomorphism from B_{2n} onto A_n, while $F|B_{2n+1}$ is a homeomorphism from B_{2n+1} onto B_{n+1}. Thus, F is a homeomorphism of M onto N.

It is not difficult to verify that $f^3 = f^2$ from which it follows that $f^n = f^2$, and thus, $\varprojlim f$ is homeomorphic to $\varprojlim f^n$ for each positive integer n.   □

4.3 Shift Mappings

If $f : [0, 1] \to [0, 1]$ is a mapping and M is the inverse limit with this single bonding map, there are homeomorphisms of M into M induced by f called *shift homeomorphisms*. One of these homeomorphisms is a shift $\overleftarrow{f}$ given by $\overleftarrow{f}(x) = (f(x_1), x_1, x_2, \dots)$; the other is its inverse $\overrightarrow{f}$ given by $\overrightarrow{f}(x) = (x_2, x_3, x_4, \dots)$. For a set-valued function f, the function $\overrightarrow{f}$ is no longer a homeomorphism, but it is a mapping of M onto M. This is our next theorem.

Theorem 4.2. *Suppose $f : [0, 1] \to 2^{[0,1]}$ is an upper semicontinuous function and $M = \varprojlim f$. Then the function $\overrightarrow{f}$ given by $\overrightarrow{f}(x) = (x_2, x_3, x_4, \dots)$ is a mapping of M onto M.*

Proof. $\overrightarrow{f}$ is the mapping that results by using $A = \mathbb{N} - \{1\}$ in Theorem 4.1. To see that $\overrightarrow{f}$ is surjective, let y be a point of M. Choose $t \in f(y_1)$ and let $x = (t, y_1, y_2, \dots)$. Then $x \in M$, and $\overrightarrow{f}(x) = y$.   □

The function $\overleftarrow{f}$ on the other hand is not a mapping but is a set-valued function; even so, there is a nonsurjective induced shift homeomorphism of this general type for set-valued functions in some cases. This is the setting for our next theorem. We actually used a homeomorphism such as the one in this theorem in Example 2.12.

Theorem 4.3. *Suppose* $f : [0, 1] \rightarrow 2^{[0,1]}$ *is an upper semicontinuous function and* $\varphi : [0, 1] \rightarrow [0, 1]$ *is a mapping of* $[0, 1]$ *into* $[0, 1]$ *such that* $\varphi \subseteq G(f)$. *If* $M = \lim \overleftarrow{f}$, *then the function* $\overleftarrow{\varphi} : M \rightarrow M$ *given by* $\overleftarrow{\varphi}(x) = (\varphi(x_1), x_1, x_2, x_3, \dots)$ *is a homeomorphism of* M *into* M.

Proof. If $x \in M$, then $\overleftarrow{\varphi}(x) \in M$ because $\varphi(x_1) \in f(x_1)$. The continuity of $\overleftarrow{\varphi}$ is a consequence of the continuity of φ. If $x, y \in M$ with $x \neq y$, it is clear that $\overleftarrow{\varphi}(x) \neq \overleftarrow{\varphi}(y)$. Thus, $\overleftarrow{\varphi}$ is continuous and 1–1 on the compact set M, so it is a homeomorphism [3, Theorem 259, p. 178].   □

We refer to the map $\overleftarrow{\varphi} : M \rightarrow M$ in Theorem 4.3 as the *shift homeomorphism induced by* φ. We revisit Example 2.12, here making use of Theorem 4.3.

Example 4.3. Let $f : [0, 1] \rightarrow 2^{[0,1]}$ be given by $f(0) = [0, 1/2] \cup \{1\}$ and $f(t) = \{t/2, 1 - t/2\}$ for $0 < t \leq 1$. Then $\lim \overleftarrow{f}$ is the union of an arc and a sequence of fans (see Figs. 2.11 and 2.12 for the graph of f and a model of the inverse limit).

Proof. Let $M = \lim \overleftarrow{f}$ and $A = \lim \overleftarrow{T^{-1}}$ be the arc lying in M where T is the full tent map, $T(t) = 2t$ for $0 \leq t \leq 1/2$, and $T(t) = 2 - 2t$ for $1/2 < t \leq 1$. Let $\varphi : [0, 1] \rightarrow [0, 1]$ be the map given by $\varphi(t) = t/2$ and $\psi : [0, 1] \rightarrow [0, 1]$ be the map given by $\psi(t) = 1 - t/2$. Let $\overleftarrow{\varphi}$ be the shift homeomorphism induced by φ and $\overleftarrow{\psi}$ be the shift homeomorphism induced by ψ.

Let $A_1 = \{x \in M \mid x_1 \in [0, 1/2] \text{ and } x_j = 0 \text{ for } j > 1\}$ and, for $n \in \mathbb{N}$, let $A_{n+1} = \overleftarrow{\varphi}(A_n)$. In the model of M in Fig. 2.12, the arc $A_i = A_{i,1}$. Let $\Gamma_0 = \bigcup_{i>0} A_i$ and note that Γ_0 is a fan with vertex $p_0 = (0, 0, 0, \dots)$ (vertices such as this one are named in reference to our model of the inverse limit). Then, $\Gamma_1 = \overleftarrow{\psi}(\Gamma_0)$ is a fan with vertex $p_1 = (1, 0, 0, \dots)$, and $\Gamma_2 = \overleftarrow{\varphi}(\Gamma_1) \cup \overleftarrow{\psi}(\Gamma_1)$ is a fan with vertex $p_2 = (1/2, 1, 0, 0, \dots)$; $\Gamma_3 = \overleftarrow{\varphi}(\Gamma_2)$ is a fan with vertex $p_3 = (1/4, 1/2, 1, 0, 0, \dots)$, and $\Gamma_4 = \overleftarrow{\psi}(\Gamma_2)$ is a fan with vertex $p_4 = (3/4, 1/2, 1, 0, 0, \dots)$. Let $\Gamma_5 = \overleftarrow{\varphi}(\Gamma_3)$, $\Gamma_6 = \overleftarrow{\varphi}(\Gamma_4)$, $\Gamma_7 = \overleftarrow{\psi}(\Gamma_3)$, and $\Gamma_8 = \overleftarrow{\psi}(\Gamma_4)$. The vertices of these fans are $(1/8, 1/4, 1/2, 1, 0, 0, \dots)$, $(3/8, 1/4, 1/2, 1, 0, 0, \dots)$, $(7/8, 3/4, 1/2, 1, 0, 0, \dots)$, and $(5/8, 3/4, 1/2, 1, 0, 0, \dots)$, respectively. Continuing in this manner, we obtain a sequence Γ of fans such that $M = A \cup (\bigcup_{i \geq 0} \Gamma_i)$.   □

For an upper semicontinuous function on $[0, 1]$, even though the map $\overrightarrow{f}$: $\varprojlim f \rightarrow \varprojlim f$ given by $\overrightarrow{f}(x) = (x_2, x_3, x_4, \dots)$ is normally not a homeomorphism, there are cases where its restriction to certain compact subsets is a homeomorphism. This is the subject of our next theorem which is a direct consequence of the theorem that a 1–1 map between compact metric spaces is a homeomorphism [3, Theorem 259, p. 178].

Theorem 4.4. *Suppose $f : [0, 1] \rightarrow 2^{[0,1]}$ is an upper semicontinuous function, $M = \varprojlim f$, and $\overrightarrow{f} : M \rightarrow M$ is the shift map given by $\overrightarrow{f}(x) = (x_2, x_3, x_4, \dots)$. If K is a compact subset of M and $\overrightarrow{f}|K$ is 1–1, then $\overrightarrow{f}|K$ is a homeomorphism.*

4.4 Other Maps Between Inverse Limits

We now turn to another theorem about mappings between inverse limits. Here we assume a sequence of maps between factor spaces in inverse limit sequences that satisfy a commutativity condition. In the case of two inverse limits with a single bonding function and a single surjective homeomorphism between corresponding factor spaces, Theorem 4.5 yields the conjugacy theorem, Theorem 2.9, as a corollary.

Theorem 4.5. *Suppose X and Y are sequences of subintervals of $[0, 1]$, and, for each positive integer i, $f_i : X_{i+1} \rightarrow 2^{X_i}$ and $g_i : Y_{i+1} \rightarrow 2^{Y_i}$ are upper semicontinuous. Suppose further, for each i, $\varphi_i : X_i \rightarrow Y_i$ is a mapping such that $\varphi_i \circ f_i = g_i \circ \varphi_{i+1}$. Then, the function $\varphi : \varprojlim f \rightarrow \varprojlim g$ given by $\varphi(x) = (\varphi_1(x_1), \varphi_2(x_2), \varphi_3(x_3), \dots)$ is continuous. Further, φ is a surjective homeomorphism if each φ_i is a surjective homeomorphism.*

Proof. The function $\Phi : \prod_{i>0} X_i \rightarrow \prod_{i>0} Y_i$ given by $\Phi(x) = (\varphi_1(x_1), \varphi_2(x_2), \varphi_3(x_3), \dots)$ is continuous and is 1–1 if each φ_i is 1–1. Because $\varphi = \Phi | \prod_{i>0} X_i$, φ inherits continuity from Φ, and it is 1–1 if each φ_i is 1–1. Thus, there are only two things to show: (1) for $x \in \varprojlim f$, $\varphi(x) \in \varprojlim g$, and (2) if each φ_i is a surjective homeomorphism and $y \in \varprojlim g$, then there is a point $x \in \varprojlim f$ such that $\varphi(x)=y$.

To show (1), we need to know $\varphi_i(x_i) \in g_i(\varphi_{i+1}(x_{i+1}))$ for each positive integer i. Suppose i is a positive integer. Because $\varphi_i \circ f_i = g_i \circ \varphi_{i+1}$, $\varphi_i(f_i(x_{i+1})) = g_i(\varphi_{i+1}(x_{i+1}))$. Because $x \in \varprojlim f$, $x_i \in f_i(x_{i+1})$ and, thus, $\varphi_i(x_i) \in \varphi_i(f_i(x_{i+1})) = g_i(\varphi_{i+1}(x_{i+1}))$. It follows that $\varphi(x) \in \varprojlim g$.

To see (2), suppose $y \in \varprojlim g$. Because each φ_i is 1–1, $x = (\varphi_1^{-1}(y_1), \varphi_2^{-1}(y_2), \varphi_3^{-1}(y_3), \dots)$ is a point of $\mathcal{Q}$ such that $\varphi(x) = y$. We now observe that $x \in \varprojlim f$. Let $i \in \mathbb{N}$. Because $g_i \circ \varphi_{i+1} = \varphi_i \circ f_i$ and $\varphi_{i+1}(x_{i+1}) = y_{i+1}$, $g_i(y_{i+1}) = \varphi_i(f_i(x_{i+1}))$. Because $y_i \in g_i(y_{i+1})$, there is a point t of $f_i(x_{i+1})$ such that $y_i = \varphi_i(t)$. But, $\varphi_i(x_i) = y_i$ and φ_i is 1–1, so $x_i = t$. Thus, $x_i \in f_i(x_{i+1})$ and we have $x \in \varprojlim f$. $\square$

The following example of Charatonik and Roe [1, Example 3.3, p. 234] shows that the condition $\varphi_i \circ f_i = g_i \circ \varphi_{i+1}$ is necessary in the hypothesis of Theorem 4.5. In Example 4.4, $\varphi_i(f_i(t)) \subseteq g_i(\varphi_{i+1}(t))$ for each positive integer i and each $t \in [0, 1]$.

Example 4.4. For each positive integer i, let $f_i : [0, 1] \to [0, 1]$, $g_i : [0, 1] \to 2^{[0,1]}$, and $\varphi_i : [0, 1] \to [0, 1]$ be given by $f_i(t) = \varphi_i(t) = t$ for each $t \in [0, 1]$, while $g_i(t) = [0, 1]$ for each $t \in [0, 1]$. Then, the map φ of Theorem 4.5 is not a surjective homeomorphism from $\varprojlim f$ onto $\varprojlim g$ because $\varprojlim f$ is an arc, while $\varprojlim g$ is the Hilbert cube.

References

1. Charatonik, W.J., Roe, R.P.: Mappings between inverse limits of continua with multivalued bonding functions. Topology Appl. **159**, 233–235 (2012)
2. Ingram, W.T., Mahavier, W.S.: Inverse limits of upper semi-continuous set valued functions. Houston J. Math. **32** 119–130 (2006)
3. Ingram, W.T., Mahavier, W.S.: Inverse limits: From continua to Chaos. In: Developments in Mathematics, vol. 25. Springer, New York (2012)

Chapter 5
Dimension

Abstract Inverse limits on $[0, 1]$ with mappings cannot raise dimension. By using set-valued functions, however, such an inverse limit can be infinite dimensional. In this chapter, we examine aspects of dimension in inverse limits on $[0, 1]$ with set-valued functions. We give an example of an inverse limit on $[0, 1]$ with set-valued functions that has dimension 2 and another having dimension 3. We conclude this chapter with a proof that an inverse limit on $[0, 1]$ with upper semicontinuous functions cannot be a 2-cell.

5.1 Introduction

In this chapter we discuss dimension of inverse limits on $[0, 1]$ with set-valued functions. In the case that the bonding functions are mappings, the dimension can never exceed 1. However, as we saw in Example 1.1, with set-valued functions the dimension of the inverse limit can be infinite.

If G is a finite collection of sets and n is a positive integer, we say that the *order* of G is n provided n is the largest of the integers i such that there are $i + 1$ members of G with a common element. If G is a finite collection of mutually exclusive sets, we say that the order of G is 0. By the *mesh* of a finite collection G of sets, we mean the largest of the diameters of the elements of G. If G and H are collections of sets, we say that H *refines* G provided for each element h of H there is an element g of G such that $h \subseteq g$. If n is a nonnegative integer, the nonempty compact metric space X is said to have dimension not greater than n, written $\dim(X) \leq n$, provided, for each positive number ε, there is a finite collection of open sets covering X that has mesh less than ε and order not greater than n. A nonempty metric space X is said to have dimension 0, written $\dim(X) = 0$, provided $\dim(X) \leq 0$. If n is a positive integer and X is a nonempty metric space, we say the dimension of X is n, written $\dim(X) = n$, provided $\dim(X) \leq n$ and $\dim(X) \nleq n - 1$. A nonempty compact metric space X is said to be *infinite dimensional* provided $\dim(X) \nleq n$ for any nonnegative integer n.

W.T. Ingram, *An Introduction to Inverse Limits with Set-valued Functions*, SpringerBriefs in Mathematics, DOI 10.1007/978-1-4614-4487-9_5, © W.T. Ingram 2012

It is convenient to use the definition of dimension given here (sometimes called *covering dimension*) in the study of inverse limits. For compact metric spaces, the property of having dimension n (respectively, not greater than n) is equivalent to the usual definition of having small inductive dimension n (respectively, not greater than n) [1, Theorem V 8, p. 67] [5, Theorem 15.2, p. 81].

5.2   Dimension 1

We begin our look at dimension in inverse limits with a well-known theorem on inverse limits on intervals with bonding functions that are surjective mappings. A proof of this theorem may be found in [4, Theorem 184, p. 127]. A similar result holds for inverse limits on $[0, 1]$ with upper semicontinuous set-valued functions whenever the bonding functions have zero-dimensional values, see Theorem 5.4 below from which Theorem 5.1 follows as a corollary.

Theorem 5.1. *If f is a sequence of surjective mappings of $[0, 1]$ onto $[0, 1]$, $\dim(\varprojlim f) = 1$.*

We now present Nall's proof of Theorem 5.3 below. First, we introduce some notation and include a couple of lemmas. Suppose $f_1, f_2, \ldots, f_n$ is a finite collection of upper semicontinuous functions such that $f_i : [0, 1] \to 2^{[0,1]}$ for $1 \le i \le n$. Recall our notation from Chap. 2 that $G'(f_1, f_2, \ldots, f_n) = \{(x_1, x_2, \ldots, x_{n+1}) \in [0, 1]^{n+1} \mid x_i \in f_i(x_{i+1}) \text{ for } 1 \le i \le n\}$ and that $G'(f_1, f_2, \ldots f_n)$ is compact by Lemma 2.1. For $n \ge 2$, let $Y = G'(f_1, \ldots, f_{n-1})$ and denote by $F_n : [0, 1] \to 2^Y$ the set-valued function given by $F_n(t) = \{(x_1, x_2, \ldots, x_n) \in Y \mid x_n \in f_n(t)\}$. This function is used extensively in this section. In [6], Nall makes the following observation.

Theorem 5.2. *Suppose $f_i : [0, 1] \to 2^{[0,1]}$ is an upper semicontinuous function for $1 \le i \le n$. Then, F_n is upper semicontinuous.*

Proof. The graph of F_n is homeomorphic to the compact set $G'(f_1, \ldots, f_n)$. By Theorem 1.2, F_n is upper semicontinuous.   □

Lemma 5.1. *Suppose $f_i : [0, 1] \to 2^{[0,1]}$ is an upper semicontinuous function for $1 \le i \le n$. If t is a point of $[0, 1]$ such that $\dim(F_n(t)) > 0$, then there exist an integer j, $1 \le j \le n$, and a point z of $[0, 1]$ such that $\dim(f_j(z)) = 1$.*

Proof. Because $F_n(t)$ is compact, if $\dim(F_n(t)) > 0$, it contains a nondegenerate continuum K, [1, Theorem D, p.22] or [5, Theorem 4.7, p.22]. Some projection of $K \subseteq [0, 1]^n$ into $[0, 1]$ is nondegenerate. Let j be the largest integer i so that $\pi_i(K)$ is nondegenerate. If $j = n$, let $z = t$ and denote by J an interval lying in $\pi_n(K)$. Then, $J \subseteq f_j(z)$ so $\dim(f_j(z)) = 1$. If $j < n$, then the projection of K into the $(j + 1)$st factor space is a single point z and, as before, $\dim(f_j(z)) = 1$.   □

Lemma 5.2. *Suppose $f_i : [0, 1] \to 2^{[0,1]}$ is an upper semicontinuous function for $1 \le i \le n$ and $\dim(f_i(t)) = 0$ for each $t \in [0, 1]$ and each i, $1 \le i \le n$. Then, $\dim(G'(f_1, f_2, \ldots, f_n)) \le 1$.*

Proof. Suppose $\varepsilon > 0$ and t is a point of $[0, 1]$. It follows from Lemma 5.1 that $\dim(F_n(t)) = 0$, so there exists a finite collection $\mathcal{V}_t$ of mutually exclusive open sets covering $F_n(t)$ such that the mesh of $\mathcal{V}_t$ is less than $\varepsilon/2$. Because F_n is upper semicontinuous, there is an open set u_t containing t of diameter less than $\varepsilon/2$ such that $F_n(u_t) \subseteq \mathcal{V}_t^*$ where $\mathcal{V}_t^*$ denotes the union of all the sets in $\mathcal{V}_t$. The collection of open sets $\mathcal{U} = \{u_t \mid t \in [0, 1]\}$ covers $[0, 1]$, so there is a finite subcollection $\mathcal{U}'$ of $\mathcal{U}$ that covers $[0, 1]$. The dimension of $[0, 1]$ is 1 so there is a finite collection $\mathcal{W}$ of open sets covering $[0, 1]$ such that the order of $\mathcal{W}$ is not greater than 1 and $\mathcal{W}$ refines $\mathcal{U}'$. Because $\mathcal{W}$ refines $\mathcal{U}'$, the mesh of $\mathcal{W}$ is less than $\varepsilon/2$. Suppose $w \in \mathcal{W}$. There is a point $x \in [0, 1]$ such that $w \subseteq u_x$ so $F_n(w) \subseteq \mathcal{V}_x^*$. Thus, $\mathcal{X} = \{v \times w \mid w \in \mathcal{W}$ and $v \in \mathcal{V}_x$ where x is a point in $[0, 1]$ such that $F_n(w) \subseteq \mathcal{V}_x^*\}$ is a collection of open sets covering $G'(f_1, f_2, \ldots, f_n)$. To see that $\mathcal{X}$ is a covering, let p denote a point of $G'(f_1, f_2, \ldots, f_n)$. There is a point $t \in [0, 1]$ such that $p_{n+1} = t$. There exist an element $w \in \mathcal{W}$ such that $t \in w$ and a point $x \in [0, 1]$ such that $F_n(w) \subseteq \mathcal{V}_x^*$. Because $(p_1, p_2, \ldots, p_n) \in F_n(t)$, $(p_1, p_2, \ldots, p_n) \in v$ for some $v \in \mathcal{V}_x$, therefore, $p \in v \times w$. If $v \times w \in \mathcal{X}$, the diameter of $v \times w$ is less than ε because the diameter of v is less than $\varepsilon/2$ as is the diameter of w. To see that the order of $\mathcal{X}$ is not greater than 1, suppose that there is a point p that belongs to three elements of $\mathcal{X}$, $v_1 \times w_1$, $v_2 \times w_2$, and $v_3 \times w_3$. Then $p_{n+1} \in w_1 \cap w_2 \cap w_3$, so some two of these are the same. Assume $w_1 = w_2$. Then, $(p_1, \ldots, p_n) \in v_1 \cap v_2$. However, there is a point $x \in [0, 1]$ such that v_1 and v_2 belong to $\mathcal{V}_x$, so $v_1 = v_2$, a contradiction. Thus, $\dim(G'(f_1, f_2, \ldots, f_n)) \le 1$. $\square$

Theorem 5.3 (Nall [6]). *Suppose f is a sequence of upper semicontinuous functions such that $\dim(f_i(t)) = 0$ for each $t \in [0, 1]$ and each integer $i \in \mathbb{N}$. Then, $\dim(\varprojlim f) \le 1$.*

Proof. Recall that $\varprojlim f = \bigcap_{n \ge 1} G_n$ where $G_n = \{x \in \mathcal{Q} \mid x_i \in f_i(x_{i+1})$ for $1 \le i \le n\}$. Observe that $G_n = G'(f_1, f_2, \ldots, f_n) \times \mathcal{Q}$. Let $\varepsilon > 0$. There is a positive integer N such that $\sum_{k \ge N} 2^{-k} < \varepsilon/2$. Let i be an integer such that $i > N$. By Lemma 5.2, $\dim(G'(f_1, f_2, \ldots, f_i)) \le 1$. Let $\mathcal{U}$ be a collection of open sets of order not greater than 1 and mesh less than $\varepsilon/2$ that covers $G'(f_1, f_2, \ldots, f_i)$. Then, $\{\pi^{-1}(u) \mid u \in \mathcal{U}\}$ is a collection of open sets of mesh less than ε and order not greater than 1 that covers $\varprojlim f$. $\square$

If $f : [0, 1] \to 2^{[0,1]}$ is a set-valued function that is the union of finitely many mappings, then f is upper semicontinuous and $\dim(f(t)) = 0$ for each $t \in [0, 1]$. Thus, our next theorem is a corollary of Theorem 5.3.

Theorem 5.4. *Suppose f is a sequence of upper semicontinuous functions such that $f_i : [0, 1] \to 2^{[0,1]}$ for each positive integer i. If the graph of f_i is the union of finitely many mappings $f_1^i, f_2^i, \ldots, f_{k_i}^i$ of $[0, 1]$ into $[0, 1]$ for each i, then*

$\dim(\varprojlim \boldsymbol{f}) \leq 1$. *Moreover, if there is a sequence $\boldsymbol{g}$ such that $g_i \in \{f_1^i, f_2^i, \ldots, f_{k_i}^i\}$ for each positive integer i and $\varprojlim \boldsymbol{g}$ is nondegenerate, then $\dim(\varprojlim \boldsymbol{f}) = 1$.*

Proof. Inasmuch as f_i is the union of finitely many mappings, $\dim(f_i(t)) = 0$ for each $t \in [0, 1]$. By Theorem 5.3, $\dim(\varprojlim \boldsymbol{f}) \leq 1$. If there is a sequence $\boldsymbol{g}$ such that $g_i \in \{f_1^i, f_2^i, \ldots, f_{k_i}^i\}$ for each positive integer i and $\varprojlim \boldsymbol{g}$ is nondegenerate, then $\varprojlim \boldsymbol{f}$ contains a nondegenerate continuum, so its dimension is 1. $\square$

Of course, one consequence of Theorem 5.4 is that inverse limits with mappings on $[0, 1]$ do not raise dimension. On the other hand, dimension may be lowered by an inverse limit on compact metric spaces even if the bonding maps are surjective as may be seen from the following example. We briefly step outside inverse limits on intervals for this example.

Example 5.1. Let φ denote the projection of the unit square $C = [0, 1] \times [0, 1]$ onto the interval $I = [0, 1]$. Let ψ denote a map of I onto C. Let $X_i = C$ and $f_i = \psi \circ \varphi$ for each i. Then, $\dim(\varprojlim \boldsymbol{f}) = 1$ even though each factor space is two-dimensional and each bonding map is surjective.

Proof. Let $Y_i = I$ and $g_i = \varphi \circ \psi$ for each i. Let $Z_i = C$ for odd integers i and $Z_i = I$ for even integers i. Let $k_i = \psi$ for odd i and $k_i = \varphi$ for even i. Using $n_1 = 1, n_2 = 3, n_3 = 5, \ldots$ in Theorem 3.1, we see that $\varprojlim \boldsymbol{k}$ is homeomorphic to $\varprojlim \boldsymbol{f}$. Using $n_1 = 2, n_2 = 4, n_3 = 6, \ldots$ in Theorem 3.1, we see that $\varprojlim \boldsymbol{k}$ is homeomorphic to $\varprojlim \boldsymbol{g}$, an inverse limit on $[0, 1]$. Therefore, $\varprojlim \boldsymbol{f}$ is homeomorphic to $\varprojlim \boldsymbol{g}$, so $\dim(\varprojlim \boldsymbol{f}) \leq 1$. Because f is surjective, by Theorem 5.1, we have $\dim(\varprojlim \boldsymbol{f}) = 1$. $\square$

By substituting an n-cell or the Hilbert cube for C in Example 5.1, we see that an inverse limit of n-dimensional or even infinite-dimensional continua can have dimension one. Inverse limits with surjective mappings on $[0, 1]$ are always one-dimensional.

5.3   Examples with Finite Dimension Greater Than 1

Our first example in this section is an inverse limit with a sequence of upper semicontinuous bonding functions on $[0, 1]$. It provides a trivial way to obtain a 2-cell as an inverse limit on $[0, 1]$.

Example 5.2. Let $\boldsymbol{f}$ be the sequence of set-valued functions such that $f_i : [0, 1] \to 2^{[0,1]}$ for each $i \in \mathbb{N}$ where $f_1(t) = [0, 1]$ for each $t \in [0, 1]$, and, for $i > 1$, $f_i(t) = t$ for each $t \in [0, 1]$. Then, $\varprojlim \boldsymbol{f}$ is homeomorphic to $[0, 1]^2$.

By replacing f_1 in Example 5.2 with the function whose graph is a given closed subset M of $[0, 1]^2$ such that the projection of M into the first factor space is surjective, it can be seen that any nondegenerate plane continuum and, indeed, virtually any compact subset of the plane is homeomorphic to an inverse limit on $[0, 1]$ using a sequence of set-valued bonding functions.

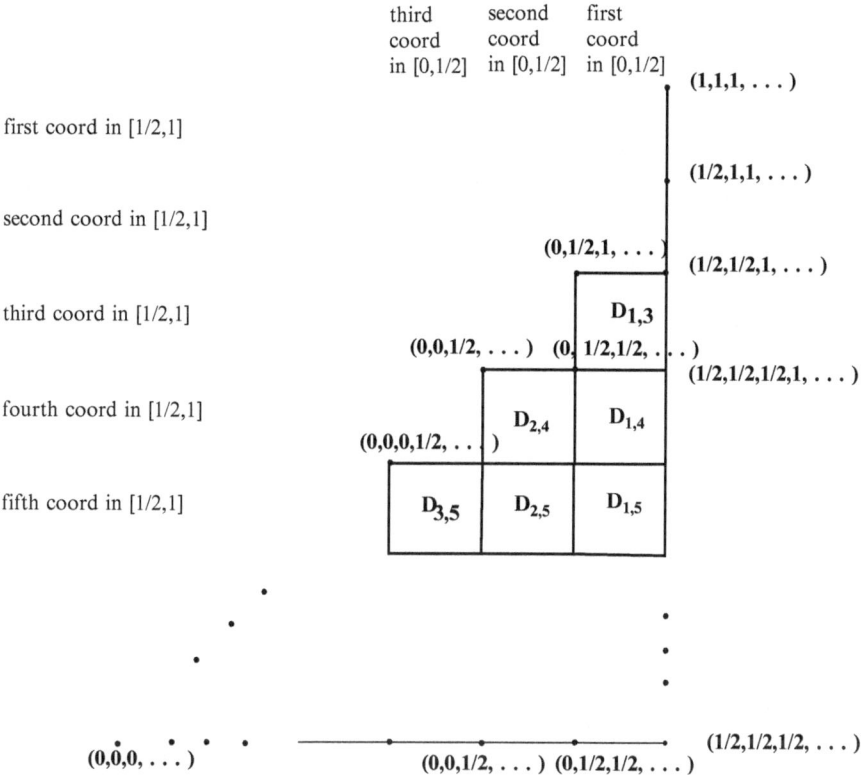

Fig. 5.1 A model of the inverse limit in Example 5.3

For inverse limits with a single surjective bonding function, the picture is quite different. We shall see in Theorem 5.5 that it is no longer possible to get even $[0, 1]^2$ as such an inverse limit. In Example 1.1, we saw that an inverse limit with a single set-valued function on $[0, 1]$ can be infinite dimensional. Such examples can also be zero-dimensional as seen in Example 1.2, and several examples we have already presented are one-dimensional even though they are obtained using only one bonding function. We now present an example from [3] of a two-dimensional continuum that is an inverse limit on $[0, 1]$ with a single bonding function and observe how to modify it to obtain examples of each finite dimension.

Example 5.3 (A 2-cell with a sticker). Let $f : [0, 1] \rightarrow C([0, 1])$ be given by $f(t) = 0$ for $0 \leq t < 1/2$, $f(1/2) = [0, 1/2]$, $f(t) = 1/2$ for $1/2 < t < 1$, and $f(1) = [1/2, 1]$. Then, the inverse limit is the union of a 2-cell and an arc intersecting the 2-cell in only one point.

Proof. Let $M = \varprojlim f$. Here M is the union of a 2-cell D and an arc A. To identify D, let i and j be positive integers with $j > i + 1$ and let $D_{i,j}$ be the 2-cell, $\{p \in M \mid p_i \in [0, 1/2], p_j \in [1/2, 1], \text{if } i > 1, \text{then } p_k = 0 \text{ for } k < i, p_k = 1/2 \text{ if } i < k < j, p_k = 1 \text{ if } k > j\}$. In Fig. 5.1, we provide a model to assist the

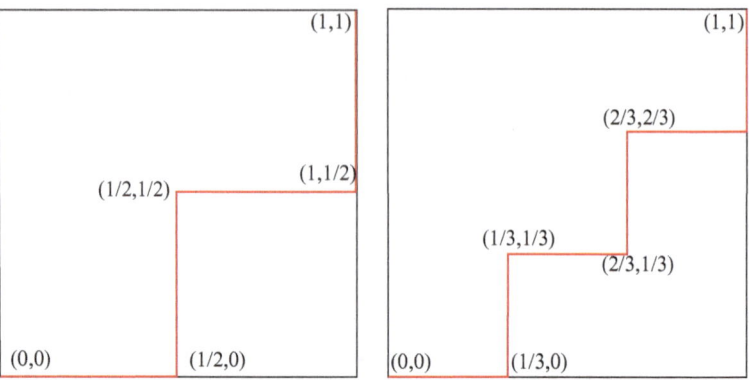

Fig. 5.2 Graphs of the bonding functions in Examples 5.3 (*left*) and 5.4 (*right*) with two-dimensional and three-dimensional inverse limits, respectively

reader. In this picture, we have labeled a few of the disks $D_{i,j}$. In the model, the disks $D_{i,j}$ and $D_{i,j+1}$ share a common horizontal border, while the disks $D_{i,j}$ and $D_{i+1,j}$ share a common vertical border as long as $i + 1 < j - 1$. Let D be the closure of the union of all the disks $D_{i,j}$ where $i \geq 1$ and $j > i + 1$.

Suppose k is a positive integer. Let $\alpha_k = \{p \in M \mid p_k \in [1/2, 1], \text{ if } k > 1, \text{ then } p_m = 1/2 \text{ for } m < k, \text{ and } p_m = 1 \text{ for } m > k\}$ and $\beta_k = \{p \in M \mid p_k \in [0, 1/2], \text{ if } k > 1, \text{ then } p_m = 0 \text{ for } m < k, \text{ and } p_m = 1/2 \text{ for } m > k\}$. Then, α_k and β_k are arcs such that α_k forms the right-hand vertical edge of the disk $D_{1,k}$ for $k = 3, 4, 5, \ldots$ in the figure, while for $k \in \mathbb{N}$, the arc β_k lies directly below all of the disks $D_{k,k+n}$ for $n = 2, 3, 4, \ldots$. The closure of $\beta_1 \cup \beta_2 \cup \beta_3 \cup \cdots$ is an arc from $(0, 0, 0, \ldots)$ to $(1/2, 1/2, 1/2, \ldots)$ forming the bottom edge of the disk D, while the closure of $\alpha_3 \cup \alpha_4 \cup \alpha_5 \cup \cdots$ is an arc from $(1/2, 1/2, 1/2, \ldots)$ to $(1/2, 1/2, 1, 1, \ldots)$ forming the right-hand edge of D.

Let $A = \alpha_1 \cup \alpha_2$. Then, A is an arc, D is a 2-cell, $M = D \cup A$, and $D \cap A = \{(1/2, 1/2, 1, 1, \ldots)\}$.   □

It is interesting to note that the inverse limit in Example 5.3 is a 2-cell with an arc attached. Theorem 5.5 below yields that it cannot be just the 2-cell. Historically, Example 5.3 preceded Theorem 5.5. We now consider the following example. We note without proof that, with obvious modifications, examples of any finite dimension may similarly be obtained.

Example 5.4 (A 3-cell with a fin). Let $f : [0, 1] \to C([0, 1])$ be given by $f(t) = 0$ for $0 \leq t < 1/3$, $f(1/3) = [0, 1/3]$, $f(t) = 1/3$ for $0 < t < 2/3$, $f(2/3) = [1/3, 2/3]$, $f(t) = 2/3$ for $2/3 < t < 1$, and $f(1) = [2/3, 1]$. Then the inverse limit is the union of a 3-cell and a fin intersecting the 3-cell in an arc. The fin is a 2-cell with a sticker where the sticker is an arc that intersects the 2-cell at only one point and the sticker misses the 3-cell entirely. (See the graph on the right of Fig. 5.2 for a picture of the graph of f.)

Proof. Let $M = \varprojlim f$, $g_1 = f \,|\, [0, 2/3]$, and $g_2 = f \,|\, [2/3, 1]$. Let $D_1 = \varprojlim g_1$ and $D_2 = \varprojlim g_2$. Suppose i, j, and k are positive integers with $j > i + 1$ and $k > j + 1$. Let $B(i, j, k) = \{x \in M \,|\, x_i \in [0, 1/3], x_j \in [1/3, 2/3], x_k \in [2/3, 1]$, and if $m \in \mathbb{N}$, then $x_m = 2/3$ if $j < m < k$, $x_m = 1/3$ if $i < m < j$, and $x_m = 0$ if $m < i$ in case $i > 1\}$. Each such $B(i, j, k)$ is a 3-cell lying in M. Then,

$$M = D_1 \cup D_2 \cup \left(\bigcup_{\substack{i > 0, j > i+1 \\ k > j+1}} B(i, j, k) \right).$$

To describe a model for this inverse limit, we first observe that D_1 and D_2 are copies of the 2-cell with a sticker from Example 5.3. To assist in identifying parts of the model, let $D_1(i, j) = \{x \in D_1 \,|\, x_i \in [0, 1/3]$ and $x_j \in [1/3, 2/3]\}$ where i and j are integers such that $j > i + 1$. Let $D_2(j, k) = \{x \in D_2 \,|\, x_j \in [1/3, 2/3]$ and $x_k \in [2/3, 1]\}$ where j and k are integers such that $k > j+1$. We use D_2 to assist in a description of a model for M; a model of D_2 is shown in Fig. 5.3. In our scheme to depict M, the set D_1 is a bit more difficult to show, but it lies at the bottom of the model. Its sticker is the arc from $(1/3, 1/3, 2/3, 2/3, 2/3, \ldots)$ to $(2/3, 2/3, 2/3, \ldots)$ at the bottom of the right-hand side of Fig. 5.3. The fin F on the 3-cell in M is the subset of D_2 that consists of the two rightmost columns of disks and the arc comprising the far right boundary of Fig. 5.3. To obtain the model for M, we build off of Fig. 5.3. In the model, we refer to "levels" in the model by the coordinate that lies in the interval $[2/3, 1]$. Under this method of viewing the model, the first 3-cell, $B(1, 3, 5)$, shows up at the fifth level down. The back face of $B(1, 3, 5)$ as shown in Fig. 5.4 is the disk $D_2(3, 5)$ shown in Fig. 5.3. The first two levels are arcs along the vertical right boundary, the third level is a disk, and the fourth level is the union of two abutting disks. In Figs. 5.4 and 5.5, we depict levels five and six. From this, we believe the reader can deduce the remaining levels without further explanation.   □

5.4 Incommensurate Continua

We now show that a 2-cell is not an inverse limit with a single upper semicontinuous bonding function from $[0, 1]$ into $2^{[0,1]}$. Recall that, unless otherwise noted, if $f : [0, 1] \to 2^{[0,1]}$ is an upper semicontinuous function, we consider the domain of the projection π_i to be the inverse limit space, $\varprojlim f$; $\overrightarrow{f}$ denotes the shift map on $\varprojlim f$ given by $\overrightarrow{f}(x) = (x_2, x_3, x_4, \ldots)$. In general, this shift map on an inverse limit with upper semicontinuous bonding functions is not a homeomorphism. However, when it is restricted to a compact set on which it is 1–1, by Theorem 4.4, its restriction is a homeomorphism. The following proof is based on a proof Nall presents in

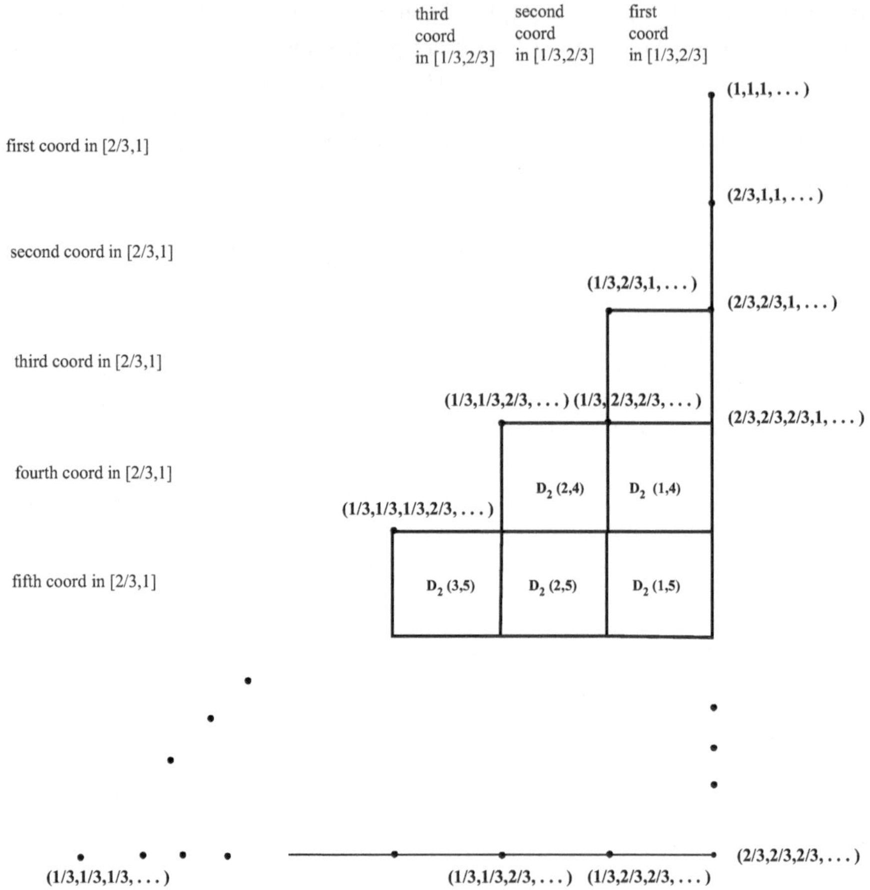

	third coord in [1/3,2/3]	second coord in [1/3,2/3]	first coord in [1/3,2/3]

$(1,1,1,\dots)$

first coord in [2/3,1]

$(2/3,1,1,\dots)$

second coord in [2/3,1]

$(1/3,2/3,1,\dots)$ $(2/3,2/3,1,\dots)$

third coord in [2/3,1]

$(1/3,1/3,2/3,\dots)$ $(1/3,2/3,2/3,\dots)$ $(2/3,2/3,2/3,1,\dots)$

fourth coord in [2/3,1]

$D_2\,(2,4)$ $D_2\,(1,4)$

$(1/3,1/3,1/3,2/3,\dots)$

fifth coord in [2/3,1]

$D_2\,(3,5)$ $D_2\,(2,5)$ $D_2\,(1,5)$

$(1/3,1/3,1/3,\dots)$ $(1/3,1/3,2/3,\dots)$ $(1/3,2/3,2/3,\dots)$ $(2/3,2/3,2/3,\dots)$

Fig. 5.3 The subset D_2 of a model of the inverse limit in Example 5.4

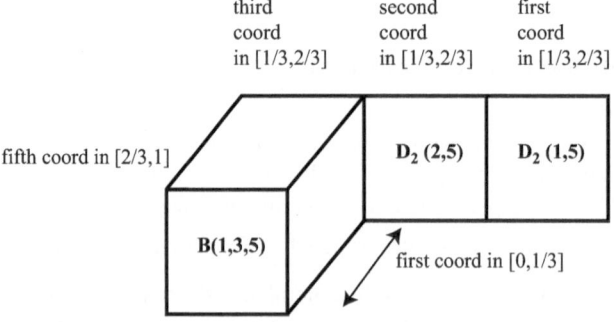

	third coord in [1/3,2/3]	second coord in [1/3,2/3]	first coord in [1/3,2/3]

fifth coord in [2/3,1]

$D_2\,(2,5)$ $D_2\,(1,5)$

B(1,3,5)

first coord in [0,1/3]

Fig. 5.4 The level five subset of a model of the inverse limit in Example 5.4

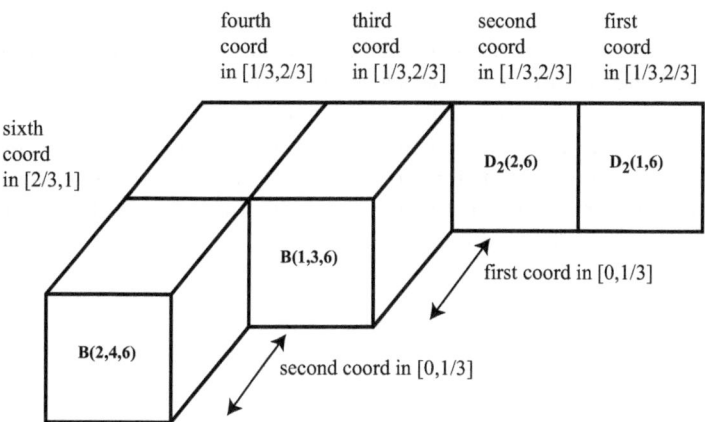

Fig. 5.5 The level six subset of a model of the inverse limit in Example 5.4

[6, Theorem 3.2, p. 1325]. In that article, he also shows, among other things, that an inverse limit with a single upper semicontinuous bonding function cannot be an n-cell for any integer $n > 1$.

Theorem 5.5 (Nall). *Suppose $f : [0, 1] \to 2^{[0,1]}$ is a surjective upper semicontinuous function. Then $\varprojlim f$ is not a 2-cell.*

Proof. Suppose $M = \varprojlim f$ is a 2-cell. If $0 < t < 1$, then $\pi_1^{-1}(t)$ separates M. Because M is a 2-cell, $\pi_1^{-1}(t)$ is not zero-dimensional [1, Corollary 2, p. 48], so it contains a nondegenerate continuum H. There is a positive integer $m \geq 2$ such that $\pi_m(H)$ is nondegenerate, but $\pi_i(H)$ is a single point for $1 \leq i < m$. Suppose J is an interval such that $\pi_m(H) = J$ and if $1 \leq i < m$, let $\pi_i(H) = \{t_i\}$ where $t_1 = t$. Let $K = \{x \in M \mid x_i = t_i$ for $1 \leq i < m$ and $x_m \in J\}$. Note that $H \subseteq K \subseteq \pi_1^{-1}(t)$.

We make use of J to show that $\pi_1^{-1}(t)$ contains an open set. By [1, Theorem IV 3, p. 44], $\pi_1^{-1}(J)$ is two-dimensional being a closed set with interior lying in a 2-cell. Let z be a point of $\pi_1^{-1}(J)$. Because $z_1 \in J$ and $J = \pi_m(H)$, there is a point w of H such that $w_m = z_1$. Let y be the point of Q such that $y_i = t_i$ for $1 \leq i < m$ and $y_{m+i} = z_{i+1}$ for $i = 0, 1, 2, \ldots$. Because $w \in H \subseteq M$ and $z \in M$, it follows that $y \in K$. Moreover, $\overrightarrow{f}^{m-1}(y) = z$. Thus, $z \in \overrightarrow{f}^{m-1}(K)$, and we have established that $\pi_1^{-1}(J) \subseteq \overrightarrow{f}^{m-1}(K)$. Because $\pi_1(K)$ is degenerate, $\overrightarrow{f}|K$ is 1–1, so $\overrightarrow{f}|K$ is a homeomorphism by Theorem 4.4. Similarly, $\overrightarrow{f}^{m-1}|K$ is a homeomorphism. Because $\overrightarrow{f}^{m-1}(K)$ contains a two-dimensional subset, K contains a two-dimensional subset. But K is a subset of $\pi_1^{-1}(t)$, so it follows that $\pi_1^{-1}(t)$ contains an open set.

Thus, we have for each t in $(0, 1)$, $\pi_1^{-1}(t)$ contains an open set. But, if $0 < s < t < 1$, then $\pi_1^{-1}(s)$ and $\pi_1^{-1}(t)$ have no point in common, so the 2-cell M contains uncountably many mutually exclusive open sets, a contradiction. $\square$

In Theorem 5.5 we saw that the 2-cell is not homeomorphic to an inverse limit on [0, 1] using a single upper semicontinuous bonding function on [0, 1]. It has been shown by Illanes [2] that a simple closed curve is not an inverse limit on [0, 1] with a single upper semicontinuous bonding function. Recently, Nall has extended this to show that the arc is the only finite graph that is an inverse limit with a single upper semicontinuous bonding function on [0, 1] [7]. It would be interesting to know which continua can be obtained as an inverse limit on [0, 1] with a single upper semicontinuous bonding function (see Problem 6.57).

References

1. Hurewicz, W., Wallman, H.: Dimension Theory. Princeton University Press, Princeton (1941)
2. Illanes, A.: A circle is not the generalized inverse limit of a subset of $[0, 1]^2$. Proc. Am. Math. Soc. **139**, 2987–2993 (2011)
3. Ingram, W.T., Mahavier, W.S.: Inverse limits of upper semi-continuous set valued functions. Houston J. Math. **32**, 119–130 (2006)
4. Ingram, W.T., Mahavier, W.S.: Inverse limits: From continua to Chaos. In: Developments in Mathematics, vol. 25. Springer, New York (2012)
5. Nadler, S.B., Jr.: Dimension Theory. Aportaciones Matemáticas, vol. 18. Sociedad Matemática Mexicana, Mexico (2002)
6. Nall, V.: Inverse limits with set valued functions. Houston J. Math. **37**(4), 1323–1332 (2011)
7. Nall, V.: The only finite graph that is an inverse limit with a set valued function on [0, 1] is an arc. Topology Appl. **159**, 733–736 (2012)

Chapter 6
Problems

Abstract This chapter contains statements of some unsolved problems in the theory of inverse limits with set-valued functions. The chapter ends with a references (current at the time of publication of this book) listing all of the books and papers on this subject that are known to the author.

6.1 Introduction

Although the main topic of this book is inverse limits on $[0, 1]$ with upper semicontinuous bonding functions, the problems listed below unless otherwise stated should be considered in compact Hausdorff spaces or perhaps compact metric spaces. Nonetheless, with each general problem, we also include a version for $[0, 1]$ because a solution in that special case is of interest.

6.2 Connectedness

Problem 6.1. Characterize connectedness of inverse limits on continua with upper semicontinuous bonding functions.

Problem 6.2. Characterize connectedness of inverse limits on continua with upper semicontinuous bonding functions on $[0, 1]$.

Problem 6.3. Find sufficient conditions that an inverse limit on continua with upper semicontinuous bonding functions be a continuum.

Problem 6.4. Solve Problem 6.3 on $[0, 1]$.

W.T. Ingram, *An Introduction to Inverse Limits with Set-valued Functions*,
SpringerBriefs in Mathematics, DOI 10.1007/978-1-4614-4487-9_6,
© W.T. Ingram 2012

Problem 6.5. Solve Problem 6.4 for upper semicontinuous functions whose graphs are unions of finitely many straight line intervals.

Problem 6.6. What can be said about compacta that are inverse limits with a single upper semicontinuous function whose graph is the union of two maps without a coincidence point?

Problem 6.7. Suppose f is a sequence of surjective upper semicontinuous functions on $[0, 1]$ and $\varprojlim f$ is connected. Let g be the sequence such that $g_i = f_i^{-1}$ for each $i \in \mathbb{N}$. Is $\varprojlim g$ connected?

6.3 The Subsequence Theorem

Problem 6.8. Find sufficient conditions on the bonding functions in the sequence f so that if $n_1, n_2, n_3, \ldots$ is an increasing sequence of positive integers, then $\varprojlim f$ is homeomorphic to $\varprojlim g$ where $g_i = f_{n_i} \circ f_{n_i+1} \circ \cdots f_{n_{i+1}-1}$.

Problem 6.9. Solve Problem 6.8 on $[0, 1]$.

Problem 6.10. Find sufficient conditions on a single bonding function so that, if n is a positive integer, then $\varprojlim f$ is homeomorphic to $\varprojlim f^n$.

Theorem 3.2 and Example 4.2 both demonstrate that for certain set-valued functions that are not mappings, it is possible for $\varprojlim f$ and $\varprojlim f^n$ to be homeomorphic.

Problem 6.11. Solve Problem 6.10 for $n = 2$.

Problem 6.12. Solve Problem 6.10 on $[0, 1]$.

Problem 6.13. Solve Problem 6.11 on $[0, 1]$.

6.4 The Closed Subset Theorem

Problem 6.14. Find sufficient conditions on the bonding functions so that closed subsets of the inverse limit are the inverse limit of their projections.

Problem 6.15. Solve Problem 6.14 on $[0, 1]$.

Problem 6.16. Solve Problem 6.14 for closed subsets of the inverse limit that are connected.

Problem 6.17. Solve Problem 6.16 on $[0, 1]$.

6.5 The Full Projection Property

Problem 6.18. Characterize the full projection property in terms of the bonding functions.

Problem 6.19. Solve Problem 6.18 on $[0, 1]$.

Problem 6.20. Solve Problem 6.18 in the case that there is only one bonding function.

Problem 6.21. Solve Problem 6.20 on $[0, 1]$.

Problem 6.22. Find sufficient conditions on the bonding functions for the inverse limit to have the full projection property.

Problem 6.23. Solve Problem 6.22 on $[0, 1]$.

Problem 6.24. Solve Problem 6.22 for inverse limits with only one bonding function.

Problem 6.25. Solve Problem 6.24 on $[0, 1]$.

Problem 6.26. Is there a general theorem about sequences of set-valued functions with the property that one of its consequences is that both Examples 3.5 and 3.6 have the full projection property?

6.6 Indecomposability

Problem 6.27. Find necessary and sufficient conditions on the bonding functions so that the inverse limit is an indecomposable continuum.

Problem 6.28. Solve Problem 6.27 on $[0, 1]$.

Problem 6.29. Solve Problem 6.27 in the case that there is only one bonding function.

Problem 6.30. Solve Problem 6.29 on $[0, 1]$.

Problem 6.31. Find necessary and sufficient conditions on the bonding functions so that the inverse limit *contains* an indecomposable continuum.

Problem 6.32. Solve Problem 6.31 on $[0, 1]$.

Problem 6.33. Solve Problem 6.31 in the case that there is only one bonding function.

Problem 6.34. Solve Problem 6.33 on $[0, 1]$.

Problem 6.35. Find sufficient conditions on the bonding functions so that the inverse limit is an indecomposable continuum.

Problem 6.36. Solve Problem 6.35 on $[0, 1]$.

Problem 6.37. Solve Problem 6.35 in the case that there is only one bonding function.

Problem 6.38. Solve Problem 6.37 on $[0, 1]$.

Problem 6.39. Find sufficient conditions on the bonding functions so that the inverse limit contains an indecomposable continuum.

Problem 6.40. Solve Problem 6.39 on $[0, 1]$.

Problem 6.41. Solve Problem 6.39 in the case that there is only one bonding function.

Problem 6.42. Solve Problem 6.41 on $[0, 1]$.

6.7   Ray with Remainder

Problem 6.43. What are sufficient conditions on a single bonding function on $[0, 1]$ so that the inverse limit is the closure of a topological ray?

6.8   Bonding Functions Vis-À-Vis Projections

Problem 6.44. Are there conditions under which $\pi_i(H) = f_i(\pi_{i+1}(H))$ for subsets H of inverse limits with set-valued functions that are not mappings?

6.9   Mapping Theorems

Problem 6.45. Find sufficient conditions that a map between inverse limits with set-valued bonding functions that is induced by monotone maps be monotone.

Problem 6.46. Replace "monotone" in Problem 6.45 by a map of your favorite type (e.g., open, confluent, weakly confluent, semiconfluent).

6.10   Hyperspaces

Segal showed that for ordinary inverse limits, the hyperspace of subcontinua of the inverse limit is homeomorphic to the inverse limit on the hyperspaces of subcontinua of the factor spaces with the induced mappings as bonding maps.

Problem 6.47. If f is a sequence of upper semicontinuous functions, does a Segal-type theorem hold?

Problem 6.48. Solve Problem 6.47 on $[0, 1]$.

6.11 Miscellaneous Problems

Problem 6.49. Find sufficient conditions on bonding functions on $[0, 1]$ so that the inverse limit is treelike.

Problem 6.50. Solve Problem 6.49 in the case of a single bonding function.

Problem 6.51. Is there a function $f : [0, 1] \rightarrow 2^{[0,1]}$ such that if $i, j \in \mathbb{N}$ and $i \neq j$, then $\varprojlim f^i$ and $\varprojlim f^j$ are not homeomorphic?

Problem 6.52. If f is a sequence of upper semicontinuous functions, what can be said about the span of $\varprojlim f$?

Problem 6.53. If f is a sequence of upper semicontinuous functions from $[0, 1]$ into $2^{[0,1]}$, what can be said about the fixed point property for $\varprojlim f$?

Problem 6.54. If f is a sequence of upper semicontinuous functions from $[0, 1]$ into $C([0, 1])$, what can be said about the fixed point property for $\varprojlim f$?

Problem 6.55. If f is a sequence of upper semicontinuous functions from $[0, 1]$ into $2^{[0,1]}$, what can be said about the fixed point property for set-valued (continuum-valued) functions on $\varprojlim f$?

Problem 6.56. If f is a sequence of upper semicontinuous functions, what can be said about the Property of Kelley in $\varprojlim f$?

Nall has shown that $[0, 1] \times [0, 1]$ cannot be obtained as an inverse limit on $[0, 1]$ with a single upper semicontinuous bonding function (although it can be obtained using a sequence of upper semicontinuous functions). Illanes showed that a simple closed curve cannot be so obtained. More recently, Nall has also shown that no finite graph other than an arc can be obtained as an inverse limit on $[0, 1]$ using a single upper semicontinuous bonding function. The examples in this book indicate that many very interesting continua can be obtained. These results suggest the following problem.

Problem 6.57. Choose a particular continuum of interest. Can it be obtained as an inverse limit with a single upper semicontinuous bonding function (on $[0, 1]$)?

Problem 6.58 (Nall). Choose the continuum in Problem 6.57 to be the dyadic solenoid (any nonchainable circle-like continuum).

Any nondegenerate plane continuum M can be obtained as an inverse limit with a sequence of upper semicontinuous bonding functions. Simply embed the continuum in $[0, 1]^2$ so that it projects onto $[0, 1]$ in the first coordinate. Because this is a closed subset of $[0, 1]^2$, it is the graph of and upper semicontinuous set-valued function. Let this function be the first term of a sequence f of functions and let all of the other terms of f be the identity on $[0, 1]$. Then $\varprojlim f$ is homeomorphic to M. (This observation has been made by almost everyone who has spent any time at all thinking about inverse limits with set-valued functions.) This observation suggests the following problem.

Problem 6.59 (Nall). Choose a nonplanar continuum of interest. Can it be obtained as an inverse limit with a sequence of upper semicontinuous bonding functions on $[0, 1]$?

Problem 6.60 (Nall). Choose the continuum in Problem 6.59 to be the two-sphere.

Some open-ended problems could be worth considering.

Problem 6.61. Choose a graph of a (simple) upper semicontinuous function on $[0, 1]$ and determine the inverse limit and a model for it with that function as a single bonding function (e.g., embed a letter of the alphabet into $[0, 1] \times [0, 1]$ so that it is the graph of an upper semicontinuous function and model its inverse limit).

Problem 6.62. In the problems listed for $[0, 1]$, replace the interval $[0, 1]$ by a simple triod (or the Cantor set, S^1, a finite tree, $[0, 1] \times [0, 1]$, or your favorite factor space).

Everything we have discussed in this book has involved inverse limit systems over the set of positive integers. In the second chapter of *Inverse Limits: From Continua to Chaos*, inverse limits with set-valued functions are discussed in a setting where the underlying directed set does not have to be the set of positive integers. For instance, it makes sense to talk about an inverse limit with set-valued functions over the set of all integers or the set of real numbers. Varagona has considered these to some extent, but the problems are generally wide open and unstated. A good starting point for the following problem would be to study inverse limits with set-valued functions on $[0, 1]$ in the case that the underlying directed set is the set of all integers or the set of all real numbers.

Problem 6.63. What can be said about inverse limits with set-valued functions if the underlying directed set is not a sequence of integers?

The following bibliography includes all of the articles on inverse limits with set-valued functions that were known to the author in April, 2012.

Bibliography

1. Banič, I.: On dimension of inverse limits with upper semicontinuous set-valued bonding functions. Topology Appl. **154**(15), 2771–2778 (2007). MR 2344740 (2008h:54020)
2. Banič, I.: Continua with kernels. Houston J. Math. **34**(1), 145–163 (2008)

3. Banič, I., Črepnjak, M., Merhar, M., Milutinović, U.: Toward the complete classification of tent maps inverse limits, preprint

4. Banič, I., Črepnjak, M., Merhar, M., Milutinović, U., Sovič, T.: Ważewski's universal dendrite as an inverse limit with one set-valued bonding function, preprint

5. Banič, I., Črepnjzk, M., Merhar, M., Milutenović, U.: Limits of inverse limits. Topology Appl. **157**(2), 439–450 (2010)

6. Banič, I., Črepnjzk, M., Milutenović, U.: Paths through inverse limits. Topology Appl. **158**, 1099–1112 (2011)

7. Charatonik, W.J., Roe, R.P.: Inverse limits of continua having trivial shape. Houston J. Math. (to appear)

8. Charatonik, W.J., Roe, R.P.: Mappings between inverse limits with multivalued bonding functions. Topology Appl. **159**, 233–235 (2012)

9. Cornelius, A.N.: Inverse limits of set-valued functions. Ph.D. thesis, Baylor University (2008)

10. Greenwood, S., Kennedy, J.: Pseudo-arcs and generalized inverse limits, preprint

11. Greenwood, S., Kennedy, J.: Connected generalized inverse limits. Topology Appl. **159**, 57–68 (2012)

12. Illanes, A.: A circle is not the generalized inverse limit of a subset of $[0, 1]^2$. Proc. Am. Math. Soc. **139**, 2987–2993 (2011)

13. Ingram, W.T.: Inverse limits and dynamical systems. In: Open Problems in Topology II, pp. 289–301. Elsevier, Amsterdam (2007)

14. Ingram, W.T.: Inverse limits of upper semi-continuous functions that are unions of mappings. Topology Proc. **34**, 17–26 (2009). MR 2476513 (2009j:54058)

15. Ingram, W.T.: Inverse limits with upper semi-continuous bonding functions: Problems and some partial solutions. Topology Proc. **36**, 353–373 (2010). MR 2646984 (2011g:54021)

16. Ingram, W.T.: Concerning nonconnected inverse limits with upper semi-continuous set-valued functions. Topology Proc. **40**, 203–214 (2011)

17. Ingram, W.T., Mahavier, W.S.: Inverse limits of upper semi-continuous set valued functions. Houston J. Math. **32**(1), 119–130 (electronic) (2006). MR 2202356 (2006i:54020)

18. Ingram, W.T., Mahavier, W.S.: Inverse limits: From continua to Chaos. In: Developments in Mathematics, vol. 25. Springer, Berlin (2012)

19. Mahavier, W.S.: Inverse limits with subsets of $[0, 1] \times [0, 1]$. Topology Appl. **141**(1–3), 225–231 (2004). MR 2058690 (2005c:54012)

20. Marlin, B.: An upper semi-continuous model for the Lorenz attractor. Topology Proc. **40**, 73–81 (2012)

21. Nall, V.: Finite graphs that are inverse limits with a set valued function on $[0, 1]$. Topology Appl. **158**(10), 1226–1233 (2011). MR 2796124

22. Nall, V.: Inverse limits with set valued functions. Houston J. Math. **37**(4), 1323–1332 (2011)

23. Nall, V.: Connected inverse limits with a set-valued function. Topology Proc. **40**, 167–177 (2012). MR 2817297

24. Nall, V.: The only finite graph that is an inverse limit with a set valued function on $[0, 1]$ is an arc. Topology Appl. **159**, 733–736 (2012)

25. Peláez, A.: Generalized inverse limits. Houston J. Math. **32**(4), 1107–1119 (electronic) (2006). MR 2268473

26. Varagona, S.: Inverse limits with upper semi-continuous bonding functions and indecomposability. Houston J. Math. **37**, 1017–1034 (2011)

27. Varagona, S.: Simple techniques for detecting decomposability or indecomposability of generalized inverse limits. Ph.D. thesis, Auburn University (2012)

28. Williams, B.R.: Indecomposability in inverse limits. Ph.D. thesis, Baylor University (2010)

Index

W.T. Ingram, *An Introduction to Inverse Limits with Set-valued Functions*,
SpringerBriefs in Mathematics, DOI 10.1007/978-1-4614-4487-9,
© W.T. Ingram 2012